Reasoning *and* Sense Making *in the* Mathematics Classroom

Pre-K–Grade 2

Michael T. Battista
Series Editor
The Ohio State University
Columbus, Ohio

NATIONAL COUNCIL OF
TEACHERS OF MATHEMATICS

Library of Congress Cataloging-in-Publication Data

Names: Battista, Michael T.
Title: Reasoning and sense making in the mathematics classroom,
  prekindergarten–grade 2 / Michael T. Battista [and four others].
Description: Reston, VA : The National Council of Teachers of Mathematics,
  [2016] | Includes bibliographical references.
Identifiers: LCCN 2016002198 (print) | LCCN 2016011166 (ebook) | ISBN
  9780873537025 (pbk.) | ISBN 9780873538770
Subjects: LCSH: Proof theory--Study and teaching (Elementary) | Logic,
  Symbolic and mathematical--Study and teaching (Elementary) |
  Mathematics--Study and teaching (Elementary) | Algebra--Study and teaching
  (Elementary) | Geometry--Study and teaching (Elementary)
Classification: LCC QA9.54 .R43 2016 (print) | LCC QA9.54 (ebook) | DDC
  372.7--dc23
LC record available at https://lccn.loc.gov/2016002198

Cover Image: ©Shutterstock/new year

The National Council of Teachers of Mathematics is the public voice of
mathematics education, supporting teachers to ensure equitable mathematics
learning of the highest quality for each and every student through vision,
leadership, professional development, and research.

The publications of the National Council of Teachers of Mathematics present a variety
of viewpoints. The views expressed or implied in this publication, unless otherwise noted,
should not be interpreted as official positions of the Council

Printed in the United States of America

Contents

Preface

The three-book series *Reasoning and Sense Making in the Mathematics Classroom* maintains the National Council of Teachers of Mathematics' (NCTM) focus on teaching that promotes and supports mathematical reasoning and sense making, and they emphasize implementation of the Common Core State Standards for Mathematics (CCSSM; NGA Center and CCSSO 2010) Standards for Mathematical Practice (SMP) and the Process Standards (PS) from NCTM's *Principles and Standards for School Mathematics* (*Principles and Standards*; NCTM 2000). To illustrate the nature of mathematical reasoning and sense making in prekindergarten–grade 8 and the critical role that reasoning and sense making play in learning and using mathematics, these books show—through student and classroom vignettes as well as instructional tasks—how instruction can support students in their development of reasoning and sense making. (All student and classroom dialogues in the books are either edited versions of actual student/classroom dialogue or composites of dialogue from research and classroom observation. Student names have been changed throughout.)

Throughout the series, research on student learning is used to help teachers understand, monitor, and guide the development of students' reasoning and sense making about core ideas in elementary school mathematics. Research on teaching and learning mathematics, as cited in the chapters, is the basis of all the discussions and recommendations in these books. To illuminate the connection between reasoning and mathematical content, all three books concentrate on sense making as it is implemented for specific content areas in prekindergarten–grade 8 mathematics learning. In this first book, targeting prekindergarten–grade 2, we focus on number and operations, early algebraic thinking, and decomposing and composing geometric shapes.

In chapter 1, Michael Battista discusses the nature of reasoning and sense making in prekindergarten–grade 2 and why they are critically important in the development of mathematical thinking. He illustrates the nature of young children's mathematical reasoning with examples of children attempting to make sense of the concepts of place value and length measurement.

In chapter 2, Arthur Baroody focuses on how early childhood instruction can engage students in mathematical reasoning while helping them to construct rich number and operation sense. He also demonstrates how reasoning builds on conceptual understanding and illustrates the important use of a learning progression to understand and guide students' reasoning and sense making.

In chapter 3, Ana Stephens, with Maria Blanton, identifies core algebraic ideas and illustrates how young children can explore and engage with these ideas in ways that lay the foundation for the future study of algebra while strengthening and deepening their existing understanding of arithmetic.

In chapter 4, Michael Battista examines children's reasoning and sense making as they decompose and compose geometric shapes—including reasoning about area—and how instruction can support the development of this reasoning. He also discusses the use of learning progressions in understanding students' reasoning and in guiding their sense making with appropriate teaching.

For your convenience in following discussions of practices and standards cited within the text, two appendixes consisting of abbreviated and labeled versions of the CCSSM Standards for Mathematical Practice and the Process Standards from NCTM's *Principles and Standards* are included in the book. There are also two appendixes of instructional tasks specific to chapter 4, "Pattern-Block Frame Tasks" and "Predict-and-Cover Tasks." You can download all four by visiting NCTM's More4U website (nctm.org/more4u). The access code can be found on the title page of this book.

Mathematical Reasoning and Sense Making[1]

Michael T. Battista

Reasoning and sense making are the foundation of mathematical competence and proficiency, and their absence leads to failure and disengagement in mathematics instruction. Thus, developing students' capabilities with reasoning and sense making should be the primary goal of mathematics instruction. In order to achieve this goal, all mathematics classes should provide ongoing opportunities for students to implement these processes.

What are mathematical reasoning and sense making? *Reasoning* is the process of manipulating and analyzing objects, representations, diagrams, symbols, or statements to draw conclusions based on evidence or assumptions. *Sense making* is the process of understanding ideas and concepts in order to correctly identify, describe, explain, and apply them. Genuine sense making makes mathematical ideas "feel" clear, logical, valid, or obvious. The moment of sense making is often signaled by exclamations such as "Aha!" "I get it!" or "Oh, I see!"

Why Focus on Reasoning and Sense Making?

Reasoning and sense making are critical in mathematics learning because students who genuinely make sense of mathematical ideas can apply them in problem solving and unfamiliar situations and can use them as a foundation for future learning. Even with mathematical skills, "[i]n order to learn skills so that they are remembered, can be applied when they are needed, and can be adjusted to solve new problems, they must be learned with understanding [i.e., they must make sense]" (Hiebert et al. 1997, p. 6).

Sense making is also important because it is an intellectually satisfying experience, and not making sense is frustrating (Hiebert et al. 1997). Students who achieve genuine understanding and sense making of mathematics are likely to stay engaged in learning it. Students who fail to understand and make sense of mathematical ideas and instead resort to rote learning will eventually experience continued failure and withdraw from mathematics learning.

Understanding How Students Think

An abundance of research describing how students learn mathematics indicates that effective mathematics instruction is based on the following three principles (Battista 2001; Bransford, Brown, and Cocking 1999; De Corte, Greer, and Verschaffel 1996; Greeno, Collins, and Resnick 1996; Hiebert and Carpenter 1992; Lester 1994; NRC 1989; Prawat 1999; Romberg 1992 Schoenfeld 1994; Steffe and Kieren 1994):

1. To genuinely understand mathematical ideas, students must construct these ideas for themselves as they intentionally try to make sense of situations; their success in constructing the meaning of new mathematical ideas is determined by their preexisting knowledge and types of reasoning and by their commitment to making personal sense of those ideas.

2. To be effective, mathematics teaching must carefully guide and support students as they attempt to construct personally meaningful mathematical ideas in the context of problem solving, inquiry, and student discussion of multiple problem-solving strategies. This sense-making and discussion approach to teaching can increase equitable student access to powerful mathematical ideas, as long as it regularly uses embedded formative assessment to determine the amount of guidance each student needs. (Some students construct ideas quite well with little guidance other than well-chosen sequences of problems; other students need more direct guidance, sometimes in the form of explicit description.)

3. To effectively guide and support students in constructing the meaning of mathematical ideas, instruction must be derived from research-based descriptions of how students develop reasoning about particular mathematical topics (such as those given in research-based learning progressions).

Consistent with this view on learning and teaching, professional recommendations and research suggest that mathematics teachers should possess extensive research-based knowledge of students' mathematical thinking (An, Kulm and Wu 2004; Carpenter and Fennema 1991; Clarke and Clarke 2004; Fennema and Franke 1992; Saxe et al. 2001; Schifter 1998; Tirosh 2000). Teachers should "be aware of learners' prior knowledge about particular topics and how that knowledge is organized and structured" (Borko and Putnam 1995, p. 42). And, because numerous researchers have found that students' development of understanding of particular mathematical ideas can be characterized in terms of developmental sequences or *learning progressions* (e.g., Battista and Clements 1996; Battista et al. 1998; Cobb and Wheatley 1988; Steffe 1992; van Hiele 1986), teachers must understand these learning progressions. They must understand "the general stages that students pass through in acquiring the concepts and procedures in the domain, the processes that are used to solve different problems at each stage, and the nature of the knowledge that underlies these processes" (Carpenter and Fennema 1991, p. 11). Research clearly shows that teacher use of such knowledge improves students' learning (Fennema and Franke 1992; Fennema et al. 1996). "There is a good deal of evidence that learning is enhanced when teachers pay attention to the knowledge and beliefs that learners bring to a learning task, use this knowledge as a starting point for new instruction, and monitor students' changing conceptions as instruction proceeds" (Bransford et al. 1999, p. 11).

Beyond understanding the development of students' mathematical reasoning, it is important to recognize that to be truly successful in learning mathematics, students must stay engaged in making personal sense of mathematical ideas. To stay engaged in mathematical sense making, students must be successful in solving *challenging but doable* problems. Such problems strike a delicate balance between involving students in the hard work of careful mathematical reasoning and having students succeed in problem solving, sense making, and learning. Keeping students successfully engaged in mathematical sense making requires us to understand each student's mathematical thinking well enough to continuously engage him or her in *successful* mathematical sense making. Furthermore, to pursue mathematical sense making during instruction, students must *believe*— based on their past experiences—that they are capable of making sense of mathematics. They must also believe that they are supposed to make sense of all the mathematical ideas discussed in their mathematics classes.

Finally, as part of the focus on reasoning and sense making in mathematics learning, students must adopt an inquiry disposition. Indeed, students learn more effectively when they adopt an active, questioning, inquiring frame of mind; such an inquiry disposition seems to be a natural characteristic of the mind's overall sense-making function (Ellis 1995; Feldman and Kalmar 1996).

Reaching All Students

The principled, student-reactive teaching described above not only helps all students maximize their learning but also benefits struggling students (Villasenor and Kepner 1993). In fact, this type of teaching supports all three tiers of Response to Intervention (RTI) instruction. For Tier 1 high-quality classroom instruction for all students, research-based instructional materials include extensive descriptions of the development of students' learning of particular mathematical topics. Research shows that teachers who understand such information about student learning teach in ways that produce greater student achievement. For Tier 2, research-based instruction enables teachers to better understand and monitor each student's mathematics learning through observation, embedded assessment, questioning, informal assessment during small-group work, and formative assessment. They can then choose instructional activities that meet their students' learning needs: whole-class tasks that benefit students at all levels or different tasks for small groups of students at the same level. For Tier 3, research-based assessments and learning progressions support student-specific instruction for struggling students so that they receive the long-term individualized instruction sequences they need.

Because extensive formative assessment is embedded in this type of teaching, support for its effectiveness also comes from research on the use of formative assessment, which indicates that formative assessment helps all students—and perhaps particularly struggling students—to produce significant learning gains, often reducing the learning gap between struggling students and their peers.

What Does Sense Making During Learning and Teaching Look Like?

The following sections explore two examples of the development of students' reasoning and sense making about particular mathematical ideas. We examine obstacles to sense making, variations in student sense making, and how teaching can support sense making at various levels of sophistication.

Making Sense of Place Value in Adding Two-Digit Numbers

To illustrate the nature of mathematical sense making and reasoning, consider how Bill, a second-grade student, approached the problem "What is 24 + 58?"

Bill: [*Writing out the traditional addition algorithm*] 8 + 4 is 12. Write the 2 here and the 1 up here. 1 + 2 + 4 = 7. Write the 7 next to the 2.

$$\begin{array}{r} 1 \\ 24 \\ +48 \\ \hline 72 \end{array}$$

Teacher: Is this really a 1?

Bill: Yes, it came from the 12; you're not allowed to put it next to the 2.

Teacher: Why not?

Bill: You're just not allowed to do that.

Bill, like the majority of young students using this algorithm, showed no evidence of making conceptual sense of this procedure. His only justification was to cite some rule that a teacher or parent had given him, which he felt compelled to follow.

In contrast, there are many ways that students can make personal sense of the problem 24 + 58, as shown in the following class discussion. Over the school year in this second-grade class, students had made sense of two-digit addition in a variety of increasingly sophisticated ways.

Teacher: All year we have been talking about addition, and you have invented a bunch of ways to add two-digit numbers. Today, I want us to take a look at all these ways. I'd like you each to solve this problem in several different ways. Then, we'll talk about what you did.

Teacher: [*After students worked on the problem 24 + 48 by themselves at their seats*] Okay, I want to hear what kinds of ways you used to solve this problem, and I want you explain to the class why you did what you did.

Fred: I remember doing the problem like this: 20, 30, 40, 50, 60, 70, 72.

Teacher: Explain what you did.

Fred: I started with the 20 in 24 and counted tens in 48. Then I counted 10 more from the 12, then 2 more.

Teacher: Where did you get 12?

Fred: It's just 4 + 8 = 12. From the 24 and 48.

Teacher: Okay, how about a different way?

Mary: I did 48, 58, 68, 70, 72.

Teacher: Explain what you did.

Mary: I started with 48, and then I did the 24. I counted 10 onto 48, then 10 more, and then 2 + 2 makes 24. If you do one part at a time, it's easy.

Teacher: Okay, how about another way?

Jon: I did 40 + 20 = 60; 8 + 4 = 12; 60 + 12 = 72.

Teacher: How did you know you could do that?

Jon: You just add the tens, then add the ones, then add them together.

Teacher: Okay, how about another different way?

Serena: I need to write it. [*Goes to board and writes as shown below, then explains*] Here's how I added: 8 + 4 is 12. So I wrote a 2 from 12 under the 8 and put the 10 from 12 over the 20, because it's tens. Then I added 10 and 20 and 40, and I got 70. And 70 + 2 is 72.

$$
\begin{array}{r}
10 \\
20 + 4 \\
+\underline{40 + 8} \\
70 + 2 = 72
\end{array}
$$

Teacher:	Why did you write the 2 under the 8?
Serena:	You just put the 2 ones from 12 in the ones place that's under the 4 and 8.

In this class discussion, we see that the students have made sense of adding 2 two-digit numbers in a wide variety of ways. In fact, the methods that these students used to solve this problem fit nicely into a *learning progression*, which is a description of the successively more sophisticated ways of reasoning and sense making that students pass through in developing a deep understanding of a mathematical idea.[2,3] Battista's learning progression for place value is outlined in chart 1.1. (See Battista 2012a for a much more detailed description, including sublevels, student examples, and suggestions for teaching students at each level.) The outline in chart 1.1 includes descriptions of the sense making and understanding of addition and subtraction algorithms that are possible for students at each learning progression level. Notice that it is not until students reach level 3 that they can even begin to truly make sense of algorithms for addition and subtraction of multi-digit numbers. Critically important is the fact that it is extremely rare for students to reach higher levels in the learning progression without first passing through the earlier levels.

No student in the previously described second-grade discussion exhibited level 5 reasoning (shown below), which we should not expect of many students until a later grade.

Jake:	I wrote it like this [*writes below on the board*]. So, $8 + 4 = 12$. That's 10, which I wrote up here, plus 2, which I wrote down here. Then $10 + 20 + 40 = 70$, and I wrote the 7 in the tens place. So the answer is 72.

$$\begin{array}{r} 10 \\ 24 \\ +48 \\ \hline 72 \end{array}$$

Teacher:	Why did you write the 7 to the left of the 2?
Jake:	I wrote 7 in the tens place because it's 70.

Premature Learning of Computational Algorithms

What happens to students' sense making when computational algorithms are taught before the students have progressed to an appropriate conceptual level in the learning progression? To see, we examine the reasoning of second grader Dion (dialogue excerpted from Battista 2012a).

Chart 1.1. Battista's (2012a) learning progression (LP) for place-value understanding of whole numbers

LP Level	Students' Conceptualization of Place Value	Possible Student Sense Making of Addition/ Subtraction Algorithms	Student Example
0	Student has difficulties counting by ones.	No sense making of place value or algorithms.	
1	Student operates on numbers as collections of ones (no skip-counting by place value).	Very little sense making of place value, and only rote use of algorithms possible.	Bill
2	Student operates on numbers by skip counting by place value (e.g., counts by tens).	Weak or no connection between place value and algorithms; only rote use of algorithms possible.	Fred Mary
3	Student operates on numbers by combining and separating place-value parts (e.g., adds tens parts without counting).	Explicit use of place value in informal multi-digit computation; emerging but incomplete understanding of place value in algorithms.	Jon
4	Student understands place value in expanded algorithms.	Place-value understanding of expanded algorithms (through hundreds).	Serena
5	Student understands place value in traditional algorithms.	Place-value understanding of traditional algorithms (through hundreds).	Jake

No student in the previously described second-grade discussion exhibited Level 5 reasoning (shown in the following dialogue), which we should not expect of many students until a later grade.

Teacher: What is the sum of 47 + 24?

Dion: [*Writing*] I added 7 and 4; it was 11. So then I put a 1 right here [*in solution*] and a 1 up here [*above 4 in 47*], and then I put 1 + 4 + 2 and it equals 7.

$$\begin{array}{r} 1 \\ 47 \\ +24 \\ \hline 71 \end{array}$$

Teacher: Okay, and then you got 71?

Dion: Yeah.

Teacher: So what does this 1 [*above the 4 in 47*] stand for?

Dion: To add with the 4 and the 2.

Teacher: Is it just a 1?

Dion: Yes.

Teacher:	Use the place-value blocks to solve the problem 36 + 28 [*shows the written problem*].
Dion:	I've got 36 and 28. [*Picks up the 2 ten-blocks from 28*] 20. [*Picks up the 3 ten-rods from 36 and lays them down*] 30. [*Points to one of the 2 ten-blocks from 28 in his hand, then lays down the 2 ten-blocks*] 40, 50. [*Places the one-blocks from both piles together and counts them, one at a time*] 51, 52, 53, 54, 55, 56, 57, 58, 59, 60, 61, 62, 63, 64. [*Writes "64" on his paper*] 64.
Teacher:	There are 25 squares under the card. How many squares are there altogether?

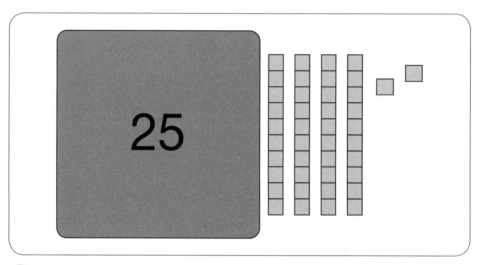

Fig. 1.1

Dion:	So, 25 squares under this [*pointing at card*], and there'd be [*pointing at columns of ten and counting by tens, then counting the last 2 squares individually*] 35, 45, 55, 65, 66, 67. 67.

We see from these examples that Dion is reasoning at level 2 *when he has place-value blocks or pictorial support*. But when he does not have visual support, he uses the standard algorithm, which he has memorized by rote (using a type of level 1 reasoning). Also, Dion does not yet combine tens and ones, reasoning that is essential for making sense of an addition algorithm.

Instruction

To determine how to encourage and support Dion to increase the sophistication of his reasoning about place value in two-digit addition, we need to examine the learning progression in chart 1.1 (Battista 2012a). According to this learning progression, our first instructional step should be to help Dion use level 2 reasoning strictly mentally, without concrete or pictorial support; that is, we

Reasoning and Sense Making in the Mathematics Classroom: Pre-K–Grade 2

want to help Dion build on the current level 2 reasoning he uses with place-value blocks and diagrams. Our second instructional step should be to help Dion move to level 3 reasoning, once again, first with visual support and then without. At this point, we should not permit Dion to use a paper-and-pencil algorithm to solve these problems.

Step 1. Moving Dion to level 2 reasoning without visual support

Teacher: How many cubes are in each pile? Write the number under the piles.

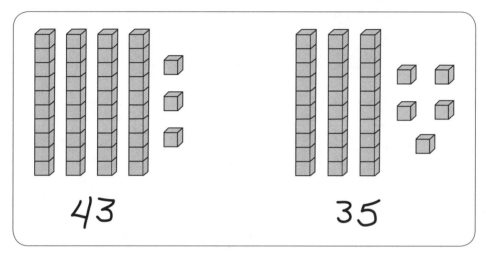

Fig. 1.2

Teacher: [*Covers the piles but not the numbers*] How many cubes are there altogether under the mats (fig. 1.3)? Can you count by tens and ones?

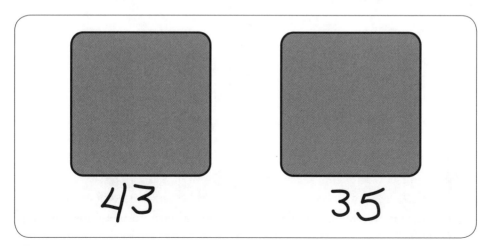

Fig. 1.3

If Dion cannot do the problem strictly mentally, without the help of visuals or pen and pencil, remove the right mat.

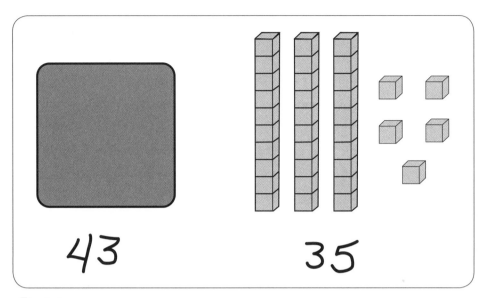

Fig. 1.4

Repeat this procedure until Dion can solve problems with both mats in place. Later present problems that are in written form but have visual support available if Dion needs it.

Step 2. Moving Dion to level 3 reasoning

Teacher: How many cubes are in each pile? Write the number under the piles. How many cubes are there altogether? Can you figure it out without counting by tens?

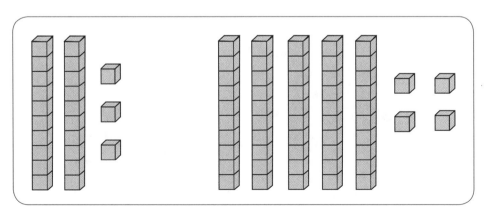

Fig. 1.5

If Dion cannot solve the problem without counting by tens, let him count by tens and then redo the problem (or do a different problem) while asking him questions that suggest and scaffold level 3 reasoning.

Teacher:	How many ten-blocks are in 23 [*pointing to blocks for 23*]?
Dion:	2.
Teacher:	How many ten-blocks are in 54 [*pointing to blocks for 54*]?
Dion:	5.
Teacher:	How many ten-blocks is that altogether? [*If Dion cannot do this addition directly, ask if counting by tens can help him, because he has shown he is able to do this.*]
Dion:	7.
Teacher:	How many one-blocks are in these 7 ten-blocks?
Dion:	70.
Teacher:	Good. Write 70 on your paper so that you remember it.
Teacher:	How many one-blocks are in 23 [*pointing to blocks for 23*]?
Dion:	3.
Teacher:	How many one-blocks are in 54 [*pointing to blocks for 54*]?
Dion:	4.
Teacher:	How many one-blocks is that altogether?
Dion:	7.
Teacher:	How many one-blocks are there altogether [*pointing across all blocks*]?
Dion:	70 and 7. That's 77.

If Dion struggles with this type of problem, use the same procedure but with the following sequence of problem types.

1. Add multiples of ten only: 30 + 50.
2. Add a multiple of ten to a mid-decade number: 34 + 50.
3. Add two mid-decade numbers, with no regrouping: 34 + 53.
4. Add two mid-decade numbers with regrouping: 36 + 58.

For each problem, ask Dion if he can add tens first. If he cannot, have him use level 2 reasoning to count by tens. Then present another problem of the same type and again ask if he can add tens first. Do not move on to a new problem type until Dion uses level 3 reasoning on the type he is currently working on.

Note how this instructional sequence, derived from a research-based learning progression, has Dion continuously building on his current reasoning, with each developmental step that he takes small enough so that his chances of completing it are very high.

Students' Reasoning and Sense Making
About the Concept of Length

To further illustrate the earlier discussion of reasoning and sense making, we examine students' reasoning about the concept of length. We look at the different ways that students are able to reason about and make sense of this topic and how instruction can encourage and support students in their development of increasingly more sophisticated levels of reasoning. There are three key steps to helping students make sense of a formal mathematical idea. First, determine empirically how they currently are making sense of the idea. Second, hypothesize how their understanding of the idea might progress. Third, choose problems and representations that can potentially help them achieve more sophisticated ways of reasoning.

The Home to School problem below (fig. 1.6) provides an excellent assessment of how well young students understand the concept of length. We first examine how students made sense of this problem, then we examine the kinds of instruction each student needs.

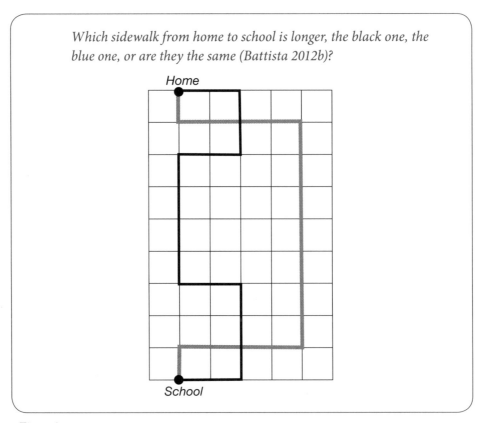

Which sidewalk from home to school is longer, the black one, the blue one, or are they the same (Battista 2012b)?

Fig. 1.6

Investigating Students' Reasoning and Sense Making about Length

Deanna says that the black sidewalk is longer because it is more "curvy." She made personal sense of the problem by relating it to her experience of walking along paths, which tends to take longer when they have many turns. (Often young students confound the length of a path with the time it takes to walk the path.)

David uses the spread between thumb and finger to draw a straight version of the blue sidewalk, one straight component at a time (right-hand drawing in fig 1.7). He uses this same procedure to construct a straight version of the black sidewalk (left-hand drawing in fig. 1.7). He then compares his drawings and says that the black sidewalk is longer. David made personal sense of the problem by straightening the paths and directly comparing them, side by side. Note that this reasoning suggests the beginnings of a valid understanding of the concept of length, and if it were done precisely (say piece by piece on a large grid with the same size units as in the problem picture), it would be mathematically correct.

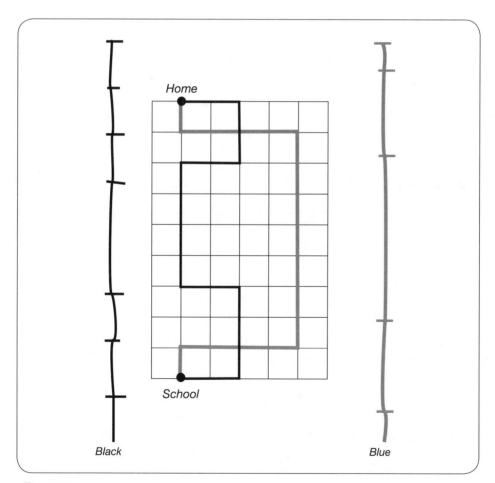

Fig. 1.7

Mollie, Matt, and Natalie make personal sense of the problem by reasoning that they should count *something*, a strategy they have often seen used in their classrooms (fig. 1.8). However, Molly and Matt do not yet understand exactly what to count. Molly counts whole (unequal) straight sections of the sidewalks and concludes that the black sidewalk is longer. Matt has observed people counting squares along paths on similar tasks, but he does not recognize how counting squares can be done in a way that corresponds to counting unit lengths; he concludes that the blue path is longer. Only Natalie correctly counts 17 unit-lengths along each sidewalk to conclude that the two paths have the same length.

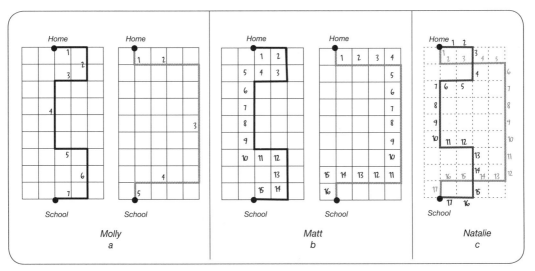

Fig. 1.8. Three students' solutions to the Home to School problem

Research Note

Difficulties in reasoning about this problem are widespread among elementary students. In individual interviews with students in grades 1–5, Battista (2010, 2012b) found that more than twice as many students used non-measurement strategies as those who used measurement strategies even though measurement strategies are most appropriate. Non-measurement strategies do not use numbers (like Deanna and David); measurement strategies use numbers (like Molly, Matt, Natalie). Furthermore, no first- or second-grade students, and only 6 percent of third graders, 12 percent of fourth graders, and 21 percent of fifth graders used correct measurement reasoning on this task (like Natalie). Even some adults have difficulty with the problem.

Instruction Focused on Individual Student Needs

Choosing instruction to help Deanna, David, Mollie, and Matt make progress in understanding length measurement requires us to understand the thinking each student can build on (see Battista 2012, for a detailed description of instructional activities). For instance, we can help Deanna by having her straighten paths and compare them directly.[4] We can help David by giving him precut sticks or rods that match the straight segment lengths on the two paths, then having him make straight paths from each set of sticks to see that the paths are the same length when carefully straightened.

We can help Molly and Matt by having them physically place unit-length rods along each path, count the unit lengths, then use the set of unit-lengths from each path to make straight versions of the paths. Because there are 17 unit-lengths in each path, when we straighten the paths, they have exactly the same length. It is important to note that most young children do not *logically* understand why counting the number of unit-lengths in the two paths tells us which is longer when straightened. They come to this conclusion *empirically* by repeatedly observing that counting predicts which path will be longer when they physically straighten the paths. It is also useful to have students compare unit-length and square iterations along grid paths to see that they generally produce different counts (unless the paths are straight). Some students make sense of this discrepancy by saying that the plastic squares have to be held "sideways"—perpendicular to the student page—to give a correct count.

The key here is to help students build on what they know to make sense of increasingly more sophisticated reasoning about length. If instruction is too abstract for students' current states of understanding, they will not be able to make the appropriate jump in personal sense making.

Integrating Conceptual and Procedural Knowledge in Reasoning About Length

Examining the conceptual and procedural knowledge needed to solve the home-to-school problem sheds additional light on student reasoning and sense making. Both types of knowledge are critical for mathematical proficiency (Kilpatrick, Swafford, and Findell 2011).

A *concept* is the meaning a person gives to objects, actions, and abstract ideas. Concepts are the building blocks of reasoning; we reason by manipulating, reflecting on, and interrelating concepts. *Conceptual knowledge* enables us to identify and define concepts, see relationships between concepts, and use concepts to reason. For instance, we might *conceptualize* length as the linear extent of an object when it is straightened or the distance you travel as you move along a path.

Procedural knowledge enables us to perform and use mathematical *procedures*, which are repeatable sequences of actions on objects, diagrams, or mathematical symbols. Procedural knowledge includes more than computational skill. For instance, to solve many length problems, students use the procedure of iterating and counting unit lengths.

The reasoning used by Molly and Matt does not properly connect conceptual and procedural knowledge. These two students have not yet developed a conceptual understanding of length measurement. Molly does not understand that the iterated units must be the same length. Matt either does not understand that the iterated units must be length units, not squares, or he thinks that iterating squares is the same as iterating unit-lengths "because they are the same size." Even more confusing for students is the fact that squares can be used to properly iterate unit-lengths for straight paths and for non-straight paths if done very carefully. Indeed, understanding the difference between iterating squares as squares and using squares to iterate unit-lengths is extremely difficult for many students. If Matt thinks that counting squares iterates unit-lengths, he may not have sufficient conceptual knowledge of unit-length iteration to regulate his use of squares to iterate unit lengths; the whole sidewalk path must be covered by a sequence of unit lengths with no gaps or overlaps in the sequence.

Summary

This discussion of the home-to-school problem illustrates that during the process of learning, students make sense of the same formal mathematical idea in different ways. For instruction to promote and support students' reasoning and sense making about length, it must be chosen to help each student build on his or her current mathematical ideas.

Observing How Instruction Can Help Students Develop the Concept of Unit-Length Iteration

We now examine how teaching can help students develop a concept of unit-length iteration that is abstract and powerful enough to deal correctly with the home-to-school problem. We first look at instruction that preceded the presentation of the home-to-school problem. Then we examine how the students made sense of this problem in light of their previously developed concepts of unit-length iteration.

Six students are working together with their teacher. The students are working on a sequence of increasingly difficult problems like the following one (fig. 1.9):

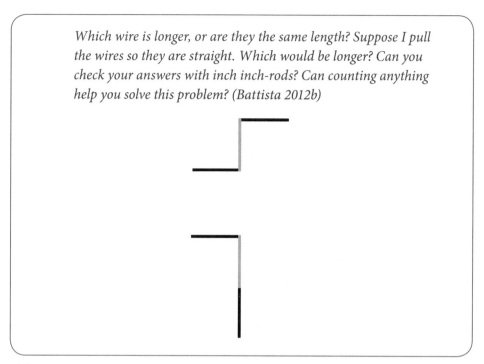

Which wire is longer, or are they the same length? Suppose I pull the wires so they are straight. Which would be longer? Can you check your answers with inch inch-rods? Can counting anything help you solve this problem? (Battista 2012b)

Fig. 1.9

The students made sense of and reasoned about this problem in a variety of ways.

Michael: I think they might be the same length because it's curved.

Kerri: [*Uses the eraser end of her pencil to move upward on the vertical segment of the bottom wire*] 1, 2, . . . 6, [*pointing at the horizontal segment of the bottom wire*] 7. [*Uses the eraser to count the 3 unit-segments on the top wire*] 1, 2, 3; 3 plus a little more. The bottom wire is longer.

Zack: Two on this [*positions pencil horizontally then slightly vertically on the bottom wire*]. And 3 on this [*positions the pencil horizontally, vertically, and then horizontally on the top wire*]. I think the top wire is longer. [*Note that this is the same type of straight-section reasoning that Molly used.*]

Brandi: Wait! I would say they're the same. Like this would be [*points at the bottom black vertical segment of the bottom wire*], like this [*points at the bottom black horizontal segment of the top wire (fig. 1.10)*]. And this line [*points at the blue vertical segment of the bottom wire*] would fit here [*points at the blue vertical segment of the top wire*], and this line [*points at the horizontal black segment of the bottom wire*] would fit here [*points at the top black horizontal segment of the top wire*].

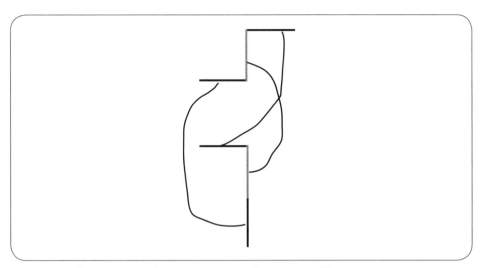

Fig. 1.10. Showing Brandi's reasoning on the wire problem

Gerald: I thought this one was longer [*top wire*], because I measured by using my fingers [*shows a finger spread*].

Kerri: The line might be an inch [*pointing at the top segment of the bottom wire*]. Then I would know that it would be 3 inches on there [*goes over the bottom wire*] and maybe 3 inches on there [*goes quickly over the top wire*]. So they're the same.

Teacher: Do you want to check with the inch-rods [*straws cut into 1-inch pieces*]? Each one of these is 1 inch long.

Kerri: 1, 2, 3 [*moves an inch-rod along the bottom wire*]. 1, 2, 3 [*moves an inch-rod along the top wire*]. They're the same length.

Teacher: Without using those inch-rods, is there anything that you could do to solve this problem by counting?

Brandi: 1, 2, 3 [*pointing to segments on the bottom wire*]. 1, 2, and 3 [*pointing to segments on the top wire*].

There is a wide variety of sophistication in students' reasoning about this problem, from Michael's non-measurement reasoning, to Kerri and Zack's incorrect counting, to Brandi's non-measurement but correct one-to-one correspondence, to Kerri's and Brandi's correct counting. To encourage and support the students in moving toward more sophisticated reasoning, the teacher not only provides students with an opportunity to check their answers, but also has them publicly discuss how they could have solved this problem by counting unit-lengths. Note, however, that some students' sense making would have benefitted from putting 3 unit-lengths on two separate wires, counting the unit-lengths, then actually straightening each wire. Also note how Brandi seemed to have made sense of Kerri's counting procedure and incorporated it into her own reasoning.

After several other problems, students are given the following problem (fig. 1.11):

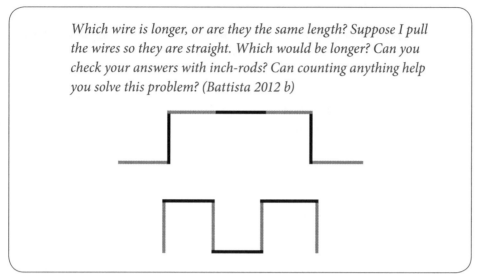

Which wire is longer, or are they the same length? Suppose I pull the wires so they are straight. Which would be longer? Can you check your answers with inch-rods? Can counting anything help you solve this problem? (Battista 2012 b)

Fig. 1.11

Zack: 1, 2, 3, 4, 5, 6, 7 [*points at segments on the top wire*]; 1, 2, 3, 4, 5, 6, 7 [*points at segments on the bottom wire*]. Both are the same.

Kerri: [*Counts 7 unit-segments on each wire.*] Yep.

Serena: [*Makes a slash on each segment on both wires as she counts (fig. 1.12a).*]

Michael: [*Writes numbers on the segments for both wires as he counts (fig. 1.12b).*]

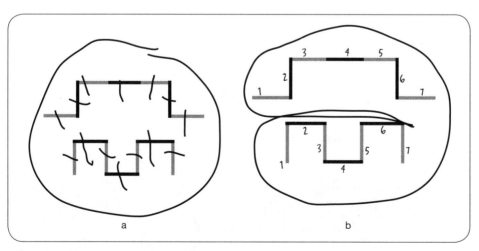

Figure 1.12

Teacher:	And do you want to check them with the inch-rods?
Kerri:	Yeah.
Students:	[*Several students at once*] No!

By the end of this session, the students were routinely counting unit-lengths to compare the lengths of the wires in problems like those shown above.

Two weeks later the teacher returns to the topic of length and has her students work on the home-to-school problem, each with his or her own two activity sheets, one path per sheet. Although the students routinely used unit-length (inch) counting on the previous set of problems, in the new context of the home-to-school problem, students abandoned this strategy. Because the sidewalk paths are drawn on square-inch grids, squares become visually salient for the students. Their concept of unit-length iteration was not abstract and general enough to apply in this new situation. The dialogue below, which took place after the students had given the strategies described above, illustrates how the students' sense making evolved.

Teacher:	When we're trying to figure out the lengths of the sidewalks, what should we count?
Gwen:	I think we should count the squares because they're like an inch.
Kerri:	The squares are as long as the segments [*points at a square along a sidewalk, then at its side*]. So they're the same length, which means that, if you chose either one of them it wouldn't be wrong because they're the same length. [*Pause*] Well, you might not come up with the same answer. 'Cause there's more squares than segments. Oh wait! Then you could just like count the squares that are nearest [*pointing at the sidewalk*].
Gwen:	And you wouldn't count ones near the corner because they're not near a segment; it's just a corner touching the line.
Kerri:	You would like want to count all the ones that have a segment on them [*points at a segment on a sidewalk path*].
Teacher:	[*Deciding that the students should all be looking at the same thing, shows Serena's sheet for the black path*] Now you're saying that this part of the sidewalk is 4 blocks [*points to the squares numbered 1–4 in fig. 1.13*], right?
Teacher:	What would happen if we were measuring this and we used our 1-inch straws?
Gwen:	That's a problem. You can't count the squares because like they would be sharing one; this and this [*pointing at the second and third unit-segments, starting at Home; each needs a square and we*

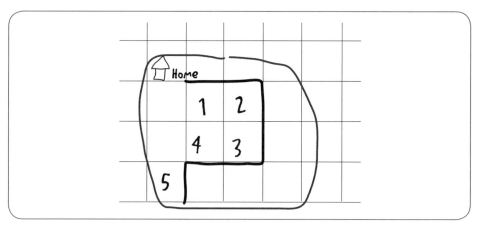

Fig. 1.13

only counted one (the 2)]. So you would need to count the sides. So we would count 1, 2, 3, 4, 5, 6, and 7 around here [*correctly counting unit-lengths on the section of the sidewalk shown in fig. 1.13*].

Teacher: So how long do you think the whole black sidewalk is?

Gwen: [*Correctly pointing to and counting unit lengths along the black sidewalk*] 1, 2, . . . , 16, 17.

Teacher: Seventeen what?

Gwen: Sides, same as our inch-straws.

Note the different ways that students made sense of this problem during the discussion. At first, Kerri thought that using squares in the grid would work because "the squares are as long as the segments." Kerri revised her reasoning and then claimed that they should count only the squares that have a side on the sidewalk path. When Serena and Gwen seemed to do what Kerri suggested (fig. 1.9), they came up with an incorrect count. So the teacher asked a question that she thought might provoke students to revise their reasoning. Gwen then realized the difficulty with counting squares; she made sense of the correct method for iterating unit-lengths along the sidewalk paths.

This episode is an excellent example of SMP 2, reason abstractly and quantitatively. The students had to decontextualize the unit-length counting strategy they used in the rod problems and re-contextualize it (transfer it) to apply it in the more difficult and complex context of the home-to-school problem. This re-contextualizing did not occur automatically; it required reasoning and sense making above and beyond the reasoning they had applied in previous problems. In fact, it was reasoning about this new problem that led students to construct a more powerful concept of unit-length iteration that could be applied to more complex situations. Most often, students' initial reasoning is context dependent; it is only by giving students a variety of contexts that students decontextualize and abstract the reasoning so that it becomes generally applicable.

Standards for Mathematical Practice and Process Standards in a Sense-Making Episode

To relate our discussion of students' reasoning about and sense making of the concept of length to the CCSSM Standards for Mathematical Practice and NCTM's Process Standards, we explicitly examine how the episodes on length are related to these practices and processes.

Standards for Mathematical Practice

Students clearly tried to make sense of the problems. Not only did the sense making differ among students but it also evolved over instructional time as students made sense of the concept of length by straightening paths, matching equal sub-lengths, and finally by ever increasingly more sophisticated counting (SMP 1a, b, g). They translated between different representations—numerical counting and spatial-unit iteration—both concretely and pictorially (SMP 1f). They made ever-increasing sense of counted quantities (SMP 2a). They constructed and evaluated arguments, and gave explanations and justifications for their work (SMP 3a, d, f). They applied the mathematics of counting and the concept of length to a real-world situation depicted in the home-to-school problem (SMP 4a). They identified important quantities and made sense of the numerical results as their notions of what must be counted evolved (SMP 4c, d). They used appropriate inch-rod tools (SMP 5). They attended to precision as they moved away from eraser estimations and toward methods of counting that were relevant to the problem, in essence creating, in action, a definition for the appropriate unit to enumerate (SMP 6b). They communicated precisely (SMP 6a). They saw structure when they used one-to-one matching of unit-segments in two wires (SMP 7)—they saw that both wires were made from the same set of linear components and thus had the same linear structure, and thus the same length. Finally, they continually evaluated their methods (SMP 8d).

Process Standards

Clearly the students built new mathematical knowledge through problem solving by implementing, discussing, and evaluating solution strategies (PS 1a, b). They reasoned and justified, developed mathematics arguments, and communicated and evaluated their thinking and strategies (PS 2a, 2c, 3a, 3b). They connected counting and spatial iteration and applied mathematics (PS 4a, b). They represented spatial unit-length iteration with counting (PS 5a, b). As shown in figure 1.14, some students even progressed from counting to reasoning using addition and fractions (PS 4a, 5b).

Kerri: I separated this one in half [*draws a vertical segment separating the bottom wire into two parts (fig. 1.14)*], and I knew 4 + 4 was 8. And the top wire is 4 + 4 [*circling the right and left sides of the top wire*] plus 1 in the middle [*pointing*]. The top is longer.

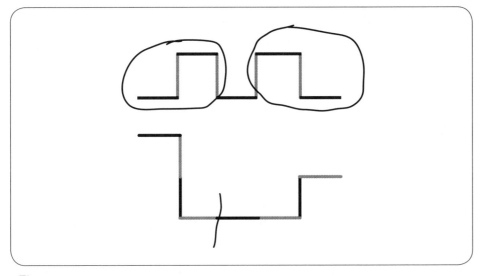

Fig. 1.14

Concluding Remarks on Mathematical Reasoning and Sense Making

To use mathematics to make sense of the world, students must first make sense of mathematics. To make sense of mathematics, students must transition from intuitive, informal reasoning stemming from their interactions with the world to precise reasoning that uses formal mathematical concepts, procedures, and symbols. The key to helping students make this transition is providing appropriate instructional tasks that target precisely those concepts and ways of reasoning that students are ready acquire. And the key to providing this support is an understanding of research-based descriptions of the development of students' increasingly more sophisticated conceptualizations and reasoning about particular mathematical concepts. Understanding the development of students' mathematical thinking is critical for selecting and creating instructional tasks, asking appropriate questions of students, guiding classroom discussions, adapting instruction to students' needs, understanding students' reasoning, assessing students' learning progress, and diagnosing and remediating students' learning difficulties.

Endnotes

1. Much of the research and development referenced in this chapter was supported in part by the National Science Foundation under Grant Numbers 0099047, 0352898, 554470, 838137, and 1119034. The opinions, findings, conclusions, and recommendations, however, are mine and do not necessarily reflect the views of the National Science Foundation.

2. An additional learning progression for the development of another aspect of number understanding is described in chapter 2, and several learning progressions for geometry are described in chapter 4.

3. It is assumed that all students pass through almost all of the levels in learning progressions. What varies is the speed at which they pass through the levels and the amount of instructional scaffolding students need to pass through each level.

4. For instruction on this task, draw the paths on square-inch grid paper and have available individual inch-rods. Also useful are sets of inch-rods strung on flexible wires so that students can make the problem paths and straighten them.

References

An, S., G. Kulm, and Z. Wu. "The Pedagogical Content Knowledge of Middle School, Mathematics Teachers in China and the U.S." *Journal of Mathematics Teacher Education* 7 (2004): 145–172.

Battista, M. T., and D. H. Clements. "Students' Understanding of Three-Dimensional Rectangular Arrays of Cubes." *Journal for Research in Mathematics Education* 27, no. 3 (1996): 258–292.

Battista, M. T., D. H. Clements, J. Arnoff, K. Battista, and C. V. A. Borrow. "Students' Spatial Structuring and Enumeration of 2D Arrays of Squares." *Journal for Research in Mathematics Education* 29, no. 5 (1998): 503–532.

Battista, M. T. "The Mathematical Miseducation of America's Youth: Ignoring Research and Scientific Study in Education." *Phi Delta Kappan* 80, no. 6 (February 1999): 424–433.

Battista, M. T. "Elementary Students' Abstraction of Conceptual and Procedural Knowledge in Reasoning About Length Measurement." Presented at the annual meeting of American Educational Research Association (AERA), Denver, Colo., April/May 2010.

Battista, Michael T. *Cognition Based Assessment and Teaching of Place Value: Building on Students' Reasoning.* Portsmouth, N.H.: Heinemann, 2012a.

Battista, Michael T. *Cognition Based Assessment and Teaching of Geometric Measurement (Length, Area, and Volume): Building on Students' Reasoning.* Portsmouth, N.H.: Heinemann, 2012b.

Borko, H., and R. T. Putnam. "Expanding a Teacher's Knowledge Base: A Cognitive Psychological Perspective on Professional Development." In *Professional Development in Education,* edited by T. R. Guskey and M. Huberman, pp. 35–65. New York: Teachers College Press, 1995.

Bransford, J. D., A. L. Brown, and R. R. Cocking. *How People Learn: Brain, Mind, Experience, and School.* Washington, D.C.: National Research Council, 1999.

Carpenter, T. P. and E. Fennema. "Research and Cognitively Guided Instruction." In *Integrating Research on Teaching and Learning Mathematics,* edited by E. Fennema, T. P. Carpenter, and S. J. Lamon, pp. 1–16. Albany, N.Y.: State University of New York Press, 1991.

Clarke, B., and D. Clarke. "A Framework for Growth Points as a Powerful Professional-Development Tool." Presented at the annual meeting of the American Educational Research Association (AERA), San Diego, Calif., 2004.

Cobb, P., and G. Wheatley. "Children's Initial Understanding of Ten." *Focus on Learning Problems in Mathematics* 10, no. 3 (1988): 1–28.

De Corte, E., B. Greer, and L. Verschaffel. "Mathematics Teaching and Learning." In *Handbook of Educational Psychology*, edited by D. C. Berliner and R. C. Calfee, pp. 491–549. New York: Simon & Schuster/Macmillan, 1996.

Ellis, R. D. *Questioning Consciousness: The Interplay of Imagery, Cognition, and Emotion in the Human Brain.* Amsterdam/Philadelphia: John Benjamins Publishing Co., 1995.

Feldman, C. F., and D. A. Kalmar. "Some Educational Implications of Genre-based Mental Models: The Interpretive Cognition of Text Understanding." In *The Handbook of Education and Human Development*, edited by D. R. Olson and N. Torrance, pp. 434–460. Oxford, U.K.: Blackwell, 1996.

Fennema, E., and M. L. Franke. "Teachers' Knowledge and Its Impact." In *Handbook of Research on Mathematics Teaching*, edited by D. A. Grouws, pp. 127–164. Reston, Va.: National Council of Teachers of Mathematics/Macmillan, 1992.

Fennema, E., T. P. Carpenter, M. L. Franke, L. Levi, V. R. Jacobs, and S. B. Empson. "A Longitudinal Study of Learning to Use Children's Thinking in Mathematics Instruction." *Journal for Research in Mathematics Education,* 27, no. 4 (1996): 403–434.

Greeno, J. G., A. M. Collins, and L. Resnick. "Cognition and Learning." In *Handbook of Educational Psychology*, edited by D. C. Berliner and R. C. Calfee, pp. 15–46. New York: Simon & Schuster/Macmillan, 1996.

Hiebert, J., and T. P. Carpenter. "Learning and Teaching with Understanding." In *Handbook of Research on Mathematics Teaching*, edited by D. A. Grouws, pp. 65–97. Reston, Va.: National Council of Teachers of Mathematics/Macmillan, 1992.

Hiebert, J., T. P. Carpenter, E. Fennema, K. C. Fuson, D. Wearne, H. Murray, A. Olivier, and P. Human. *Making Sense: Teaching and Learning Mathematics with Understanding.* Portsmouth, N.H.: Heinemann, 1997.

Kilpatrick, J., J. Swafford, and B. Findell. *Adding It Up: Helping Children Learn Mathematics.* Washington, D.C.: National Academy Press, 2011.

Lester, F. K. "Musing About Mathematical Problem-Solving Research: 1970–1994." *Journal for Research in Mathematics Education* 25, no. 6 (1994): 660–675.

National Council of Teachers of Mathematics (NCTM). *Principles and Standards for School Mathematics.* Reston, Va: NCTM, 2000.

National Governors Association Center for Best Practices (NGA Center) and Council of Chief State School Officers (CCSSO). *Common Core State Standards for Mathematics. Common Core State Standards (College- and Career-Readiness Standards and K–12 Standards in English Language Arts and Math).* Washington, D.C.: NGA Center and CCSSO, 2010. http://www.corestandards.org/wp-content /uploads/Math_Standards1.pdf

National Research Council. *Everybody Counts.* Washington, D.C.: National Academy Press, 1989.

Prawat, R. S. "Dewey, Peirce, and the Learning Paradox." *American Educational Research Journal*, 36, no. 1 (1999): 47–76.

Romberg, T. A. "Further Thoughts on the Standards: A Reaction to Apple." *Journal for Research in Mathematics Education,* 23, no. 5 (1992): 432–37.

Saxe, G. B., M. Gearhart, and N. S. Nasir. "Enhancing Students' Understanding of Mathematics: A Study of Three Contrasting Approaches to Professional Support." *Journal of Mathematics Teacher Education* 4 (2001): 55–79.

Schifter, D. "Learning Mathematics for Teaching: From the Teachers' Seminar to the Classroom." *Journal for Mathematics Teacher Education*, 1, no. 1 (1998): 55–87.

Schoenfeld, A. C. "What Do We Know About Mathematics Curricula?" *Journal of Mathematical Behavior* 13 (1994): 55–80.

Steffe, L. P. "Schemes of Action and Operation Involving Composite Units." *Learning and Individual Differences*, 4, no. 3 (1992): 259–309.

Steffe, L. P., and T. Kieren. "Radical Constructivism and Mathematics Education." *Journal for Research in Mathematics Education* 25, no. 6 (1994): 711–733.

Tirosh, D. "Enhancing Prospective Teachers' Knowledge of Children's Conceptions: The Case of Division of Fractions." *Journal for Research in Mathematics Education*, 31(1): 5–25.

van Hiele, P. M. *Structure and Insight*. Orlando, Fla.: Academic Press, 1996.

Villasenor, A., and H. S. Kepner. (1993). "Arithmetic from a Problem-Solving Perspective: An Urban Implementation." *Journal for Research in Mathematics Education* 24 (1993): 62–70.

Using Number and Arithmetic Instruction as a Basis for Fostering Mathematical Reasoning

Arthur J. Baroody

> *Asked what the largest number was, Nikki, a kindergartner,*
> *responded, "A million." The next question—"What number do*
> *you think comes after a million?"—was clearly disconcerting.*
> *Nevertheless, after a moment's thought, the girl responded, "A*
> *million and one." Asked what she thought came after a million*
> *one, Nikki again thought for a moment and then answered,*
> *"A million and two." Pressed further by a follow-up question—*
> *"What comes after a million two?"—she reflected thoughtfully*
> *for a few moments and concluded, "There is no largest number."*

● ● ●

Although Piaget suggested that young children are seriously limited in their reasoning ability, in fact, as the case of Nikki illustrates (Baroody 2005), they are capable of powerful reasoning when provided with adequate support (see, e.g., Ennis 1969; Evans 1982; Harnett and Gelman 1998; Morris 2000; Smith 2002, 2003, 2008). The focus of this chapter is on how teachers can adapt early-childhood instruction to support students' mathematical reasoning while helping them to construct a rich number and operation sense. Addressed first are the various types of mathematical reasoning Nikki used to construct the concept of *indefinite number succession* (an example of the relatively abstract concept of *infinity*). The second section illustrates how reasoning builds on conceptual understanding, and underscores the importance of using a learning progression. The third section shows how early number and arithmetic instruction can serve as a basis for fostering the various forms of mathematical reasoning.

Connection to Common Core State Standards for Mathematical Practices and Principles and Standards for School Mathematics Process Standards

In this chapter we illustrate how students use the following Standards for Mathematical Practice (SMP): making sense of problems by analyzing relationships (SMP 1b); reasoning abstractly and quantitatively by making sense of relations among quantities (SMP 2a) and generalizing beyond particular contexts or cases (SMP 2b); constructing viable arguments by making conjectures (SMP 3b), recognizing counterexamples (SMP 3c), justifying conclusions (SMP 3d), and reasoning inductively (SMP 3e); modeling with mathematics by solving real-world problems (SMP 4a) and reflecting on whether or not a result makes sense (SMP 4d); using appropriate tools strategically by employing mathematical knowledge to detect possible errors (SMP 5c); attending to precision by communicating precisely to others (SMP 6a), using clear definitions to reason (SMP 6b), and calculating fluently (SMP 6c); looking for and making use of structure by looking for patterns (SMP 7a); and looking for and expressing regularity in repeated reasoning by looking for general methods and shortcuts (SMP 8a), noticing regularities (SMP 8b), and overseeing the problem-solving process while attending to the details (SMP 8c).

The chapter also illustrates the following Process Standards (PS): solving problems by using a variety of strategies (PS 1a) and building new mathematical knowledge through problem solving (PS 1b); using various types of reasoning and proof (PS 2a), making and investigating conjectures (PS 2b), and developing and evaluating mathematical arguments (PS 2c); clearly communicating, analyzing, and evaluating mathematical thinking and strategies (PS 3a, b, c); and interconnecting, applying, and representing mathematical ideas (PS 4a, b; 5a).

Types of Mathematical Reasoning

Nikki applied four types of mathematical reasoning used by mathematicians, if only in an informal way.

Intuitive Reasoning

Definition

Intuitive reasoning involves using what you know to make an educated guess. Specifically, it entails playing a hunch, that is, using what is obvious (appearances), what feels right (an assumption), or previous experience (precedent) to draw a conclusion. For example, when asked what number comes after a million, Nikki may have assumed from her previous experience that when adults ask a question it indicates her answer (a million is the largest number) was incorrect or that the teacher would not have asked about the number after a million unless there was one.

Caveats

Although Nikki's educated guess (There "must" be a number larger than a million) turns out to be true, there is no guarantee that intuitive reasoning—which is based on appearance, assumptions, or precedent—will result in a correct conclusion. After all, appearances can be deceiving (e.g., the line that looks to be the shortest may not have fewer people in it); assumptions can be wrong (e.g., taller does not always mean older); and precedent does not ensure future trends (e.g., a question from an adult does not always indicate a given answer is incorrect).

Uses

Even so, conclusions from intuitive reasoning can provide a useful starting point for mathematical inquiry or problem solving. Intuitive reasoning can be the source of conjectures—possible explanations that can be checked by collecting evidence or using other forms of reasoning—or can suggest possible strategies for solving a problem.

Development

Young children are undoubtedly capable of using intuitive reasoning.

Empirical Induction ("Patterning")

Definition

Inductive reasoning can be thought of informally as "bottom-up" reasoning; that is, inducing or discovering a pattern, relation, or regularity entails examining particular cases (examples) and discerning a commonality among them (discovering a generality). For example, from the succession of questions (examples), Nikki recognized a pattern, namely that even the "largest" numbers she could conceive each have a number after them (inducing the generality that every number has a successor).

Caveats

The commonality Nikki discovered from considering three examples happens to be one that can be established by a mathematical or logical proof. However, young children and even adults make the common error of assuming that a pattern found among the few examples under scrutiny must be true of all possible cases. For example, examining only three elements of a number sequence—3, 5, 7—might lead to the induction that the sequence must be the odd numbers: 1, 3, 5, 7, 9, 11. . . . However, this conclusion is wrong if it turns out that the first number in the sequence is 2, not 1, in which case the sequence is the prime numbers: 2, 3, 5, 7, 11, 13. . . .

Uses

As with intuitive reasoning, conclusions derived from inductive reasoning can provide a useful starting point for mathematical inquiry or problem solving by serving as a source of conjectures.

Development

Humans are born pattern detectors. Finding a commonality is the heart of concept formation, and even nonverbal infants begin to construct concepts by detecting patterns in adult talk or in nonverbal experiences.

Deductive ("Logical") Reasoning

Definition

Informally, deductive reasoning can be thought of as top-down reasoning. It involves reasoning from a generality or generalities (premise or premises) assumed to be true in order to draw a conclusion about a particular case. For instance, if all even numbers are divisible by 2 (premise 1) and we know that 3,571,938 is divisible by 2, then we conclude that 3,571,938 is an even number. When asked "What comes after a million (or million one)?", Nikki probably used her knowledge of the repeating patterns in the counting sequence to respond. Specifically, she may have reasoned that, after a term starting a new series (e.g., a decade term such as *twenty* or a hundred term such as *three hundred*), the rest of the series is generated by combining this first term with the single-digit sequence *one* to *nine* (e.g., after *twenty* comes *twenty-one, twenty two,* . . . ; after *three hundred* comes *three hundred one, three hundred two,* etc.—premise 1) and, if a *million* introduces a new series (premise 2), then the next terms **must** be a *million one,* a *million two,* and so forth.

Caveats

If a premise is not true or the logic is not valid, the conclusion may or may not be true. For example, if the shortest distance between two points is a straight line (premise 1) and points A and B are on the surface of the earth (premise 2), is the shortest air route from A to B a straight line (conclusion)? No, because premise 1 is true for a flat surface, not a sphere's surface.

Uses

Deductive reasoning involves applying logic to what is known to draw a conclusion about what is unknown. Unlike intuitive and inductive reasoning, a conclusion drawn by deduction, such as Nikki's, *necessarily* follows from what is given; it is *necessarily* true (*if* the premises are true and the logic is valid).

Development

Inhelder and Piaget (1964) suggested that children younger than seven years of age or so were preoperational thinkers incapable of logical (deductive) reasoning and that children about seven to twelve years of age were capable of reasoning logically about concrete (familiar) situations but not about abstract ones. However, evidence indicates that children younger than seven years of age can reason logically at least about familiar, if not abstract, situations (Donaldson 1978). Consider an example of a kindergarten teacher using fictitious names unfamiliar to her students to present a problem involving transitive reasoning: logical reasoning of ordered items (e.g., If A > B and B > C, then it follows that A is also > C).

Teacher:	If Raja is older than Jordi and Jordi is older than Liam, who is older, Raja or Liam?
Mei:	Jordi.
Asha:	Raja.
Dora:	You didn't tell us that.
Uma:	Not fair.
Teacher:	[*After checking a list of her students' birthdays, restating the problem using familiar names*] Dora is older than Mei, and Mei is older than Uma, so who's the oldest?
Dora:	[*Immediately*] I'm the oldest. [*Mei and Uma nod in agreement.*]

One problem with unfamiliar situations is that young children must make an effort to remember all the information given (e.g., the unfamiliar names of fictitious children, such as Raja or Liam, and both stated relations: Raja is older than Jordi, and Jordi is older than Liam). However, in situations where they can readily remember the given information (e.g., familiar names and relations that are personally meaningfully), children can be successful on simple transitive-reasoning tasks (Bryant and Trabasso 1971).

Mathematical Induction

Definition

Similar to empirical induction, mathematical induction entails drawing a generalization about a reoccurring pattern, often after considering multiple examples. Unlike empirical induction, the domain of mathematical induction is restricted to the natural (counting) numbers, and, like deductive reasoning, the generalization is necessarily true (of all subsequent counting numbers). Whereas the essence of informal mathematical induction is exploration, the focus of (formal) mathematical induction is proving a generalization. With informal mathematical induction, the justification for a generalization may only be implicit or informal. For example, Nikki recognized that a million was not the largest number, because the cognitive conflict and resulting reflection due to her teacher's probe prompted her to conclude that it has a successor (a specific next number). She then dismissed a million one as the largest number, because the follow-up probe impelled her to conclude this example also had a number after it. In response to yet another follow-up probe ("What comes after a million one?"), she apparently realized that the process of determining the next number in the counting sequence would not end and concluded, *There is no largest number.* Note that the teacher did not spur Nikki to justify her conclusion or explain her reasoning, and so the logic of her argument remained entirely implicit.

Caveats

Informal mathematical induction works in the special case of reasoning about the positive integers, by which it is understood that n is followed by $n + 1$, which in turn, is followed by $(n + 1) + 1$, and so forth.

Uses

Like deductive reasoning, then, informal mathematical induction is creative in that it allows one to use observations and existing knowledge to create new knowledge, that is, to extend understanding beyond the limits of existing knowledge. As the case of Nikki illustrates, it can be an important process in young children's meaningful mathematical learning. Indeed, it even enabled Nikki to comprehend what she could not experience directly (the infinite).

Development

Intuitively, it might seem that Nikki is an exceptional case and that it is not reasonable to expect most five-year-olds to engage in a form of reasoning as sophisticated as informal mathematical induction or to derive concepts as abstract as the indefinite succession principle. What very little research exists seems to indicate that five- to seven-year-olds are capable of informal (if sometimes fallible) mathematical induction (Smith 2008). Moreover, some research indicates that primary-level pupils can comprehend the idea of indefinite succession (Evans 1983; Evans and Gelman 1982). Hartnett and Gelman (1998), for example, found that the majority of their grade 2 and a quarter of their kindergarten participants understood that every natural number has a successor, and half of the kindergarten pupils were classified as "waverers" (responded inconsistently).

A Learning Progression That Underscores the Interrelated Development of Number Sense and Reasoning

In actuality, the growth of number sense—the network of how numbers behave and their relations—and reasoning competence go hand in hand. On one hand (as the case of Nikki illustrates), reasoning can contribute new insights or ideas that expand number sense. On the other hand, all forms of reasoning in young children, even logical reasoning, build on experience or knowledge. The more extensive children's number sense, the better they can make educated mathematical guesses (conjectures), the more likely they are to see mathematical patterns (particularly less obvious ones), and the more probable they will be able think mathematically in a logical manner. For example, it is difficult to troubleshoot systematically and logically a car, television, or computer problem if you do not understand how such contraptions work.

Following is a discussion of a possible learning progression that includes how students like Nikki can arrive at the point where they can successfully engage in informal mathematical induction and construct a concept of indefinite number succession. Exhibit 2.1 summarizes the learning progression. After discussing the developmental components of the learning progression, the instructional implications of such a progression are considered.

Developmental Components of the Learning Progression

The progression details the aspects of number sense necessary to engage in informal mathematical induction and how other forms of reasoning helped provide this developmental basis. Note that the steps in this possible learning progression start with children's first number words, which typically appear at about the age of two years. Embedded in the learning progression commentary are specific instructional activities that can be used to encourage students to move to a next level in the learning progression.

Cardinal Concepts and Verbal Subitizing of Small Numbers.

Young children initially do not understand that the number word *one* represents a single item, *two* represents a pair of items, *three* represents a trio of items, and so forth. For example, two-year-old Arianne indicated that two, one, three, five, and ten fingers were all "two." As with many other concepts, children may construct number concepts via an *inductive process* involving examples and counterexamples. As children see a number word associated with various visual or tactile examples (e.g., "*two* eyes," "*two* hands," "*two* socks," "*two* shoes," "*two* cars") but not with counterexamples (e.g., "take *two* cookies, not *three* cookies," "that's *five* fingers, not *two*"), they gradually construct exact cardinal concepts of

Exhibit 2.1. A learning progression for early number development

1. *Cardinal concept of small numbers + verbal subitizing* (**1 and 2 first, then 3, and— in time—4 to 6**)	By seeing different examples of a number labeled with a unique number word and counterexamples labeled with other number words, children construct precise cardinal concepts one, two, and three (Palmer and Baroody, 2011). For example, seeing various pairs (e.g., 🖐🖐, ⭕⭕, ■■, 🎴🎴) labeled "two" can help a child recognize that this number word refers to number (as opposed to a particular shape, color, or other feature irrelevant to number) and multiple items at that (as opposed to a singular item), and counterexamples ("take one cookie, not two") can help a child understand that "two" refers only to pairs. These concepts permit **verbal subitizing**: The ability to immediately recognize and label small collections with an appropriate number word.
2. *Ordinal concept of small numbers* (**collections of one to about three items**)	**Verbal subitizing** enables children to see that "two is *more* than one" item and that "three is *more* than two" items (understand the term *more*) and that numbers have an **ordinal meaning**.
3. *Meaningful object counting* (**includes the cardinality principle: last count word indicates the total number of items**)	**Verbal subitizing** enables children to understand the principles underlying **meaningful counting**: *stable order, one-to-one*, and *cardinality principles*. For example, by watching an adult count a small collection that a child can recognize as "three," she or he can understand why the last number word in the count is emphasized or repeated: it represents the total or how many (the cardinal value of the collection).
4. *Increasing magnitude principle + counting-based number comparisons* (**especially collections larger than three**)	**Verbal subitizing** and **ordinal number concept** lead to discovery of the **increasing magnitude principle**: the counting sequence represents increasingly larger quantities. This enables children to use **meaningful object counting** to determine the larger of two collections (e.g., seven items is more than six items because you have to count further to get to *seven* than you do for *six*).
5. *Mental comparisons of non-neighboring or non-successive numbers*	**Familiarity with the counting sequence** and **the increasing magnitude concept** provides a basis for comparing two numbers that are at obviously different positions in the counting sequence (i.e., to make gross comparisons of 2 or 7, 10 or 3, 9 or 5, and 4 or 8).
6. *Number-after knowledge* of the counting sequence	Familiarity with the counting sequence enables a child to enter the sequence at any point and *specify the next number* instead of always counting from one.

Exhibit 2.1. (Continued)

7. *Mental comparisons of neighboring or successive numbers* ("Number after" = more)	The use of the *increasing magnitude principle* and *number-after knowledge* enables children to determine efficiently and *mentally compare even close numbers*, such as the larger of two neighboring numbers (e.g., "Which is more, seven or eight? Eight.").
8. *Successor Principle* ("Number after" = 1 more)	*Verbal subitizing* enables children to see that "two" is exactly one more than "one" item and that "three" is exactly one more than "two" items, knowledge that can help them understand the *successor principle*: each successive number in the counting sequence is exactly one more than the previous number.
9. *Reconceptualization of the counting sequence as the (positive) integer sequence*	The *successor principle* enables children to view the counting sequence as $n, n + 1, [n + 1] + 1, \ldots$ (*positive integer sequence*) as a linear representation of number.
10. *Informal mathematical induction*	An understanding of the *successor principle* and a linear representation of the *(positive) integer sequence* provide a basis for informal mathematical induction.
11. *Infinite succession principle (concept of infinity)*	*Informal mathematical induction* enables children to use their number-after knowledge and the successor principle to realize that the positive integer sequence can, in principle, go on forever.

Note: This exhibit is an adaptation and expansion of table 3 in Frye et al. (2013).

small numbers: first with one and two, later with three, and even later with four. Activity 1 illustrates how helping children construct exact cardinal concepts of small numbers and engaging them in reasoning can go hand in hand.

Exact cardinal concepts of small numbers provide the conceptual basis for verbal subitizing, immediately and reliably recognizing the total number of items in small collections and labeling them with an appropriate number word (Kaufman et al. 1949). Exact cardinal concepts and verbal subitizing of small numbers is a key basis for the meaningful learning of a variety of number, counting, and arithmetic concepts and skills (see Baroody, Lai, and Mix 2006 for a detailed discussion). Activity 2 illustrates how helping children master verbal subitizing of small numbers and engaging them in reasoning can go hand in hand.

Activity 1. The "Can You Find?" Game

Aim: Provide an opportunity to identify examples and counterexamples of small numbers (SMP 1b, 3c, 4d, 5c) and construct exact understandings of *one* and *two*, next *three*, and then *four* by empirical induction (SMP 2a, 2b, 3e; PS 1b, 2a).

Procedure: With a small group of children, or the whole class, encourage participants (one at a time) to identify an example of "one thing" in the classroom or in a children's story. The other participants can either "agree" or "disagree." Require voters who disagree to justify their vote by pointing out a second instance of the item.

Teacher:	Jorge, now it's your turn to find something in the room of which is there is only one.
Jorge:	One turtle.
Teacher:	Class, do you agree or not?
Class:	[*Unanimously*] Agree!
Teacher:	Silvia, your turn to find something in the room of which is there is only one.
Silvia:	One flag.
Teacher:	Class, do you agree or not?
Class:	[*In near accord*] Agree!
Virginia:	Disagree. There's not just one flag. There is another [*the state*] flag in the closet.

Parenthetically, note that Virginia's objection raises the issue of how to define *room*—the constraints on the space that could be searched and, thus, what constitutes a valid answer. Silvia is correct IF *classroom* is defined narrowly as the observable portion of the room; Virginia is correct IF it is defined more broadly to include the closet and other closed areas, such as cabinets. Considering constraints explicitly cannot begin too early. For example, many children believe that "addition always makes bigger." Although this generalization is true for adding the counting numbers, it is not true for broader ranges of numbers (e.g., when adding zero and another number or when an addend is a negative integer). As noted in the next paragraph, encouraging children to consider counterexamples can be an invaluable means of defining the constraints that apply to a generalization.

Along with finding examples of *one*, have participants try to find examples of *two*. Teaching *one* and *two* together underscores the differences between the terms: a single instance versus multiple instances. In other words, a constraint for the general idea of *one* is that it does not include multiple instances, and a constraint for the general concept of *two* is that it does not include single instances. Once children can reliably identify *one*, add *three* to the mix. This helps define the upper boundary of *two* (i.e., the constraint of where *two* ends and "more than *two*" or "many" begins). Once children can reliably identify *two*, focus on *three* and add *four* to the mix. This helps define the upper boundary of *three*, that is, what is "more than *three*." As with *one*, children who disagree about an example of *two* or *three* should be asked to justify why.

Activity 2. The "Hidden Fingers" Game

Aim: Help children associate *one* with a single finger, *two* with two fingers, and *three* with three fingers and, once mastered, *four* with four fingers and *five* with five fingers (PS 1b). Engage children in using the process of elimination, a form of logical thinking (e.g., If it's not one then it must be two or three [PS 1a, 2a]). Help children recognize that even an incorrect answer can be helpful in solving a problem.

Procedure: The game can be played with a child or a team of two children. Explain: "I will hold up one finger (show one finger and have the player/s do the same), two fingers (show two fingers and have the player/s do the same), *or* three fingers (show three fingers and have the player/s do the same) behind my back. You have three chances to guess how many fingers I am hiding behind my back. Ready? How many fingers do you think I am holding up now?" If the player responds by holding up fingers, ask how many fingers he or she is holding up. If the child cannot respond, identify the number of fingers for the child. If the child is correct, say "Yes, I was holding up *n* fingers behind my back" and show the number of fingers. If the guess is incorrect, say: "No, I am not hiding *n* fingers; guess again." If a child guesses a number more than *three*, say, remind the child that you are holding up one finger, two fingers, or three fingers. For children who struggle, provide them with a clue, such as "Here's a clue; I am holding up two fingers. Show me with your fingers how many fingers I am hiding." Once children can reliably subitize *one* to *three*, play an advanced version of the game in which the three choices are three, four, or five fingers.

Ordinal Meaning of Small Numbers

Initially, young children do not realize that number words can represent the relative size of collections. Verbal subitizing enables children to see that "two is *more* than one" item and that "three is more than two" items, construct an ordinal (as well as a cardinal) meaning of numbers, and understand the comparative meaning of the term *more*. Activity 3 illustrates how helping children construct an ordinal meaning of small numbers and engaging them in reasoning can go hand in hand.

Activity 3. The Advanced "Hidden Fingers" Game

Aim: Help children to understand the meaning of *more* and *fewer* and to recognize *three* as more than *two* and *two* as more than *one* or, in other words, to recognize that numbers can be ordered in terms of relative magnitude (PS 1b). Engage children in using the process of elimination, a form of logical thinking (e.g., If it's not *one* then it must be *two* or *three*; PS 1a, 2a). Help children recognize that reflecting on incorrect answers can be helpful in solving a problem.

Procedure: The game is played in the same manner as activity 2, except that a player gets only two guesses but receives feedback. For instance, if two fingers are hidden and a player guesses "one," the feedback would be "No, I am hiding more than one finger." If the player guessed "three," the feedback would be "No, I am hiding fewer than three fingers."

Rule-Governed Verbal Counting: Key Aspect and Developmental Prerequisite of Meaningful Object Counting

Verbal counting is often (inaccurately) referred to as "rote counting" and is frequently taught in rote fashion (as a skill that must be memorized by rote via repetitive practice). In fact, although the first ten numbers in the English count are entirely arbitrary symbols that must be memorized by rote, the counting sequence beyond ten has many patterns that can be discovered via empirical induction and that also can serve as rules and the basis of meaningful memorization. For example, except for *eleven* and *twelve*, which come from the old German "one over ten" and "two over ten," respectively, and two variations of the pattern/rule (*thirteen* and *fifteen*), the teens follow a simple pattern: single-digit number word + *teen* (e.g., six + teen, seven + teen, eight + teen). Activity 4 illustrates how a focus on discovering these patterns and their exceptions can serve as a basis for meaningfully memorizing the counting sequence.

Activity 4. "Cora the Confused Counter" (Baroody and Coslick 1998)

Aim: An error-detection game can help children induce or recognize regularities in the verbal counting sequence and their exceptions and provide an opportunity to discuss them (SMP 3c, 5c, 6a, 7a, 8b; PS 1b).

Procedure: A teacher can pretend to take on the identity of Cora the Confused Counter or assign a doll the identity and ask the class to help Cora learn how to count. Specify that the class or group can help Cora by pointing out when she counts correctly and when she counts incorrectly and—in both cases—why. Modeling examples of correct counting can serve as opportunities for children to discover regularities. Modeling examples of incorrect counting can serve as opportunities for children to bring exceptions to rules to their attention. The following are some examples of incorrect counts:

1. Count 1, 2, 3, 4, 5, 1, 2, 3, 4, 5 or 1, 2, 3, 3, 4, 5. (Error: You are not allowed to repeat numbers; each number must be unique.)

2. Count 1, 4, 2, 5, 3. (Error: You are not allowed to count in any order; counting words must be used in the same order.)

3. Count 1, 2, . . . , 12, three-teen or 1, 2, . . . , 14, five-teen (exceptions to the teen rule).

4. Count 1, 2, . . . , nine-teen, ten-teen, eleven-teen. (Nine as in 19 or 29 indicates the end of one decade and the need for a new decade term.)

Meaningful Object Counting

Young children often learn to count collections in a one-to-one fashion through imitation and memorization by rote. As a result, they do not understand, for instance, why adults emphasize or repeat the last number word in the counting process. Some children may even imitate this behavior but not realize that the last number word indicates the total (cardinal value of a collection). Verbal subitizing and intuitive or inductive reasoning enable children to understand the "whys" as well as the "hows" of object counting (instead of learning this enumeration skill through imitation or memorization by rote). For example, by watching an adult count a small collection that a child can recognize as *three*, the child may better understand why the adult emphasizes "three" or says "see three" after completing the count, especially if this process is repeated with examples of *three* and with examples of other small numbers; that is, the child may infer that counting is just another way of determining the total of a collection. This would involve intuitive reasoning if the child made the inference after seeing a single example, or inductive reasoning if the child saw the commonality over several examples. In any case, such experiences may prompt children to induce the *cardinality principle of counting*, that is, to *understand* that the last number word in the count is emphasized or repeated because it represents the *total* or how many (the cardinal value of the collection). Activity 5 illustrates how helping children construct the cardinality principle of counting and engaging them in reasoning can go hand in hand.

Activity 5. The "Hidden Blocks Clue" Game

Aim: Provide an opportunity to discover the *cardinality principle of counting* via empirical induction with small collections and to generalize the principle with larger collections (SMP 2a, 2b, 3e, 7a; PS 1b).

Procedure: The game can be played with a child or a small group of children. First place a collection of two to four items (within the players' subitizing range) behind a screen and explain: "I am hiding some blocks behind this screen. You have one chance to guess how many blocks there are. Here's a clue." Count the now hidden blocks out loud, emphasizing the last number word used in the process (the cardinal value of the collection). Encourage a player to make a guess.

If there are multiple players, each player can be given interlocking blocks or other objects to show his/her guess. Alternatively, each player can whisper his/her guess in the teacher's ear or record it on a mini slate board using tallies or numerals. After players have indicated their guess, reveal the previously hidden collection. After children are successful playing the game with small collections, play the game with collections of five to seven items. Note that checking guesses after the blocks are revealed can serve to model or practice one-to-one counting.

Increasing Magnitude Principle of Counting and Counting-Based Number Comparisons

Between three and one-half and six years of age, children undergo critically important changes in their thinking about the counting numbers from one to ten. By four years of age, children generalize their ordinal concept of small numbers to their knowledge of the counting sequence. For example, they may infer (via empirical induction) that if two is more than one and appears later in the counting sequence and that the same relation is true for three as well as two, then *any* number that comes later in the counting sequence is larger than earlier numbers (Schaeffer, Eggleston, and Scott 1974). The increasing magnitude principle enables children to make concrete counting-based number comparisons, that is, to determine the larger of two collections by counting them (Sarnecka and Carey 2008). The collection that requires the longer count is "more." Activities 6 and 7 help children construct or apply the increasing magnitude principle; activity 8 illustrates how learning to use the counting sequence to compare the size of collections can go hand in hand with reasoning.

Activity 6. The "Disappearing Numbers" Game (Baroody 1989)

Aim: Provide an opportunity to discover that the counting sequence represents increasingly large quantities (the increasing magnitude principle) via empirical induction with small collections and then larger collections (SMP 2a, 2b, 3e, 7a; PS 1b).

Procedure: The game can be played with one adult and one child or with two or three children alternating turns. When playing with one child, allow the child to go first. With two or three children, roll a die or use some other random process to determine who goes first.

In the basic version, lay out number squares 1 to 5 on a table face up and in numerical order. In the advanced version, lay out the number squares 1 to 10 in the same manner. Above each number square, have a child or several children represent the number with interlocking cubes (see fig. 2.1). After building a "number staircase" with the interlocking cubes, have the children count as you point to each number square in order of magnitude. This will reinforce the counting sequence and help identify each numeral on a number square with its equivalent number word and quantity (the number of interlocking cubes).

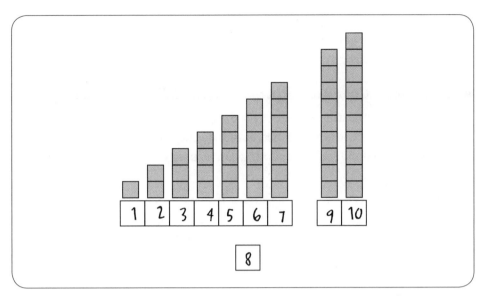

Fig. 2.1. A number staircase for the advanced "Disappearing Numbers" game

Give a player a choice of two numbers and ask which is bigger. Whether playing the basic or advanced version, initially use extreme comparisons, that is, those involving a number that a child can subitize and another number that is obviously larger. As the child succeeds, use less extreme comparisons. If a prompt is needed, ask the child (or, if necessary, help the child) to identify the relative positions of the numbers on the number staircase model. If a player can say which is more, that child wins the number square with the larger number. For example, in fig. 2.1 the child indicated "eight" (was more than two) and so took the number square eight. The eight interlocking cubes can be left in place or—as shown above—can be removed by the teacher to make the game somewhat more challenging. Make sure to alternate the order of the smaller and larger term so that the larger number is not always presented last (e.g., "Which is more, one or five?" "Which is more, five or two?" "Which is more, eight or two?" "Which is more, three or seven?").

If incorrect, encourage the child (or, if need be, help the child) to select the cubes representing each number in the number staircase and directly compare them. For example, for *seven* or *three*, ask the child to pick up the seven interlocking cubes and then the three interlocking cubes; place the three stack next to the seven stack and choose the stack with more. For a child who continues to struggle, use an entire staircase and illustrate the increasing magnitude principle with several extreme examples that include a number the child can subitize. For example, say: "Let's see whether *two* or *ten* is more. Let's find *two* on the number staircase." Encourage (or, if necessary, help) the child to count the cubes and emphasize the two numbers involved in the comparison, "One, **two**, three, four, five, six, seven, eight, nine, **ten**." For some children, it might help to follow up with questions such as "Which is just a few, two or ten?" "Which is a lot, two or ten?" Note that the idea is to help the child discover the increasing magnitude principle for him- or herself so that he or she "owns it."

Activity 7. Basic Comparisons of the "Hidden Collections" Game

Aim: Provide an opportunity to discover the *increasing magnitude principle* via empirical induction with small collections and generalize the principle to larger collections (understanding that counting can be used to compare the size of two collections; SMP 2a, 2b, 3e, 7a; PS 1b).

Procedure: The procedure is similar to that in activity 5, except that two collections—a row of red blocks and a row of blue blocks—are each counted out of view behind a vertical screen, and a player has to decide which is larger based on the length of the count. Begin with small collections of four and one, one and three, two and one, and two and four. Note that the order in which the smaller and larger collections are counted should be varied so that it requires a player to listen carefully and not simply always choose the second count. For example, announce "I have some red blocks and some blue blocks behind the screen. I am going to count the red blocks: one, *t-w-o*. Now I am going to count the blue blocks: one, two, three, *f-o-u-r*. Are there more red blocks or more blue blocks? After participants have made their guesses, reveal the two collections and confirm which row of blocks is larger by counting the smaller first ("One, two") and the larger second, emphasizing the fact that you are surpassing the first count: "One, *two*, three, *FOUR*." Once children catch on, use two collections larger than four that are not close in size (e.g., five and eight, nine and six, or ten and seven). Note that judging the larger of the two collections requires logical (deductive) reasoning; the larger collection requires a longer count (premise 1). Four blocks requires a longer count than two blocks (premise 2). So, four blocks is more than two blocks.

Activity 8. Cards "More Than" (Wynroth 1986)

Aim: Provide an opportunity to discover that counting (as well as subitizing) can be used to compare collections and to practice the use of counting to determine the larger collection (SMP 2a, 2b, 3e, 7a; PS 1b).

Procedure: The game is similar to the card game War, except that it is played with a deck of cards with pictured collections and children are encouraged to count to determine whose card has the larger or largest collection. In the basic version of the game, use cards depicting a collection of one to five. Initially, have children play in pairs and encourage them to count (even though they can subitize the collection). The following exchange occurred while a teacher had Matías and Sofia demonstrate how to play the game to others and used the opportunity to help students discover how counting could be used to compare collections.

Matías:	[*After drawing a card from his deck with* ✳△◆] 3.
Sofía:	[*After drawing a card from her deck with* ■●✶✶■] A lot; a lot beats 3.
Teacher:	How could we use counting to determine which card has more items?
Sofía:	Count them.
Teacher:	Show me.
Sofía:	[*Counts Matías's card*] 1, 2, 3. [*Then counts her card.*] 1, 2, 3, 4, 5.
Teacher:	So how does that help you to know which is more?
Matías:	Sofía has to use more numbers [*number words to count her collection*].
Teacher:	What do you mean?
Matías:	Sofía counted all the way to 5.
Mariana:	See [*counts Sofía's card*] 1, 2, 3—that's what Matías has; [*continuing the count using Sofía's card*] and 4, 5—that's what Sofía has.
Teacher:	[*Clarifying*] Matías counted, "1, 2, 3," and Sofía counted, "1, 2, **3**, and then 4, **5**."

With the advanced version, use collections of one to ten with up to four players.

Teacher:	Who wins the cards this round?
Tandy:	Me.
Violet:	Me.
Teacher:	Well, let's see. Tandy, you count the objects on your card.
Tandy:	1, 2, 3, 4, 5, 6.
Teacher:	Violet, you count the objects on your card.
Violet:	1, 2, 3, 4, 5, 6, 7, 8.
Teacher:	So, who has more?
Violet:	Me, me, me; I counted far [*beyond 6*].
Teacher:	[*Clarifying*] Violet counted past 6 all the way to 8: [*while recounting the objects on Violet's card*] 1, 2, 3, 4, 5, **6**, 7, **8**.

Mental Comparisons of Non-Neighboring Numbers

The increasing magnitude principle also allows children to make abstract or mental number comparisons, that is, to deduce which of two spoken or written numbers is larger. Children can first use this principle and their familiarity with the counting sequence to determine which of two numbers far apart in the counting sequence is larger. Activities 9 and 10 illustrate how learning to use the counting sequence to make mental comparisons of numbers can go hand in hand with reasoning.

Activity 9. Basic and Advanced "Number-List Race" Game
(Baroody, 1989)

Aim: Provide an opportunity to practice using the *increasing magnitude principle* and logic (deductive reasoning) to compare the relative size represented by two stated and non-consecutive number words (PS 1a, 1b).

Procedure: Two or three children or a child and an adult can play. In the basic version a player picks a dot card from a deck, turns it over, and subitizes or counts the number of dots on the card, then moves his or her car that number of spaces on the racetrack. As illustrated in fig. 2.2, Yellow-Blossom turned over the five-dot card, counted it, and then counted five spaces to move her car to 5 on the number list track. Subsequently, Zov moved to 4 and Xia moved to 8.

Teacher:	So whose race car moved furthest?
Xia:	Mine!
Teacher:	Zov—the third-place winner—only counted up to 4, Yellow-Blossom—the second-place winner—counted up to 5, and Xia—the first-place winner—counted all the way to 8.

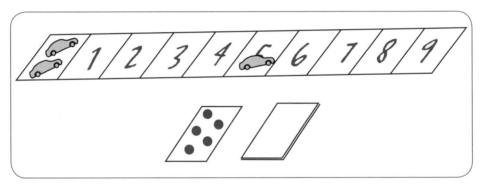

Fig. 2.2

In the advanced (reasoning-based) Number-List Race, a child's turn starts with him or her turning his or her back to the number-list racetrack and covering his or her eyes so that the number list cannot be used as an aid in determining the larger number. The teacher or another player draws a card from a deck of cards on which two numbers are printed, such as 5 or 1, 7 or 4, 2 or 6, 3 or 8. The teacher or another player announces the number comparison indicated on the card (e.g., "Which is more, five or one?"). The player must decide which number is larger. With the number-list racetrack out of view, the child is required to use the increasing magnitude principle as a premise and deduce which of the announced numbers is larger.

Activity 10. The "Guess My Number" Game

Aim: Provide an opportunity to practice using the *increasing magnitude principle* and the terms *more than* and *less than* and using the process of elimination (a form of logical or deductive reasoning) to determine a number (PS 1a, b).

Procedure: Two to four children can play. One way to play is determine who goes first, second, and so forth by some random process. The moderator chooses a number from one to five or, in the advanced version, from six to ten and announces, "I have chosen a number from one to five (six to ten). The first player to guess the number wins a point. The first player then makes his or her guess. If correct, the player is awarded a point in the form of a block, tally, or other representation. If incorrect, the moderator provides feedback. For instance, if the moderator chose four, and the child guessed three, the moderator would say "No, my number is more than three." Note that players can use this feedback and knowledge of the counting sequence (e.g., the increasing magnitude principle) to deduce that one and two are not options and that four and five are. If the moderator chose four, and the child guessed one, then the moderator would specify "No, my number is more than one." The next player would be given a chance to guess the number. This process would be repeated until the number is correctly guessed. The next round would start with the child who guessed correctly. Children may reason that going first in this game does not put one in the best position to win.

Number-After Knowledge

If children are asked to identify which number comes after another, initially they have to start with *one*, count up to the given number, and then count once more to determine the number after it. For example, when asked "What number comes after nine when we count?" three-year-old Alison was stumped. When asked "What comes after "1, 2, 3, . . . , 7, 8, 9," she quickly responded "10." As children become familiar with the counting sequence, they no longer need a "running start" to determine the number that follows another. Instead they can access the counting sequence at the given number and state its successor. This skill is critical for the (efficient) application of the next two steps in the learning progression. As activity 11 illustrates, a game can provide opportunities for children to practice number-after relations and engage in logical (deductive) reasoning.

Activity 11. Dominoes "Just After" (Wynroth 1986)

Aim: Provide an opportunity to practice number-after relations and—using these relations as premises—involve children in logical (deductive) reasoning (PS 1a, b).

Procedure: The game is identical to dominoes, except that an added domino must have a side that is one more than—"is the number after"—an open side of a domino already placed (fig. 2.3). To start the game, the dominoes are evenly distributed to the players.

Teacher: Who has the double 2 to start the game?

Andrea: I do [*places the double 2 on the board*].

Teacher: Markus, it's your turn. Do you have a domino with the number just after two?

Markus: What's that?

Teacher: Let's count together: 1, 2 [*pointing to the dots on the left side of Andrea's domino as they count*], . . . , and then comes?

Markus: 3.

Teacher: Do you have a domino that has a three side Markus?

Markus: [*Examines his dominoes and places the 3-4 domino with its three side next to the right-hand side of Andrea's double-2 domino*] Yep.

Teacher: Acqwon, it's your turn. Do you have a play?

Using the game rules (match an open end with the next number), the existing game board (the two open ends available are two and four), and his number-after knowledge (after two is three and after four is five) as premises, Acqwon deduces that he can play the 5-3 domino he has. As shown below, Acqwon places the five end of the domino next to the open four end of Markus's domino.

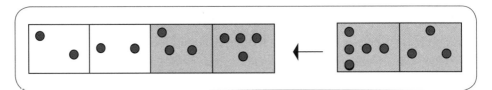

Fig. 2.3

Mental Comparisons of Neighboring Numbers

Even if children know the increasing magnitude principle, they may not know the number-after relations (well enough) to compare neighboring numbers such as seven and eight fluently. Once children become fluent with number-after relations, they can use this knowledge and the increasing magnitude principle as well as fluent number-after knowledge to deduce the larger of two neighboring numbers efficiently (see, e.g., activity 12).

Activity 12. Racer's Choice

Aim: Provide an opportunity to use the increasing magnitude principle and known number-after relations to deduce the larger of two neighboring numbers and to justify the conclusion logically (SMP 3d, 6a; PS 1a, b).

Procedure: The game is similar to the Advanced Number List Game, except that a (more abstract) non-numbered circular track and comparison of consecutive numbers are used.

Teacher:	[*After drawing a card on which 6 or 5 is printed*] Troy, would you rather move your race car 6 spaces or 5 spaces? Which is more, 6 or 5, and why?
Troy:	6, because it comes after 5 when we count.
Teacher:	So, why does that make 6 more than 5.
Troy:	The further you count, the bigger the number.
Teacher:	[*After drawing a card on which 4 or 5 is printed*] Maureen, would you rather move your race car 4 spaces or 5 spaces? Which is more 4 or 5—and why?
Maureen:	5, because it's further from 1 [the first counting number] and the numbers keep getting bigger [as you count further and further from 1].
Teacher:	[*After drawing a card on which 6 or 7 is printed*] Ferdinand, would you rather move your race car 6 spaces or 7 spaces? Which is more 6 or 7, and why?
Ferdinand:	I'd rather go 10.
Teacher:	That's not an option. 6 or 7?
Ferdinand:	7, because you have to count 6 and then one more [number word] to get to 7.

Note that children can use the same line of reasoning to determine the larger of even larger numbers. For example, if asked which is larger, nineteen or eighteen, a child can logically reason that as the former comes after the latter in the counting sequence and the later numbers in the sequence are greater than earlier ones, so nineteen *must* be more than eighteen.

Successor Principle and Reconceptualization of the Counting Sequence as the Integer Sequence

Although children may know that eight is more than seven, it does not guarantee that they realize a number is exactly one more than its preceding number neighbor and that this is true for all number neighbors. In time, preschoolers use small-number subitizing to see that *two* is exactly one more than *one* item and that *three* is exactly one more than *two* items. By generalizing this induced pattern to the counting sequence as a whole, children construct the *successor principle*: "Each successive number name refers to a quantity that is [exactly] one larger" than its predecessor in the counting sequence (Common Core State Standards 2010, K.CC.B.4c). Helping young children construct the successor principle itself can involve engaging children in mathematical reasoning. Recognizing that each successive number in the counting sequence is exactly one more than its predecessor is typically children's first introduction to a *growing number pattern*. Unlike repeating patterns such as or AABAABAAB, which have a core that repeats (e.g., AB in ABABAB patterns or AAB in AABAABAAB patterns), growing patterns get larger (or smaller) by a constant amount. The counting sequence, for instance, involves a constant +1 function. The purpose of Activities 13 and 14 is to help children recognize this +1 function.

Activity 13. Birthday Candles

Aim: Use a child's birthday as a pretext for adding candles to a birthday cupcake, relating successive numbers (number-after relations) to age (number of birthdays) and using empirical induction to discover that each successive number (birthday) is represented by the addition of a single candle (unit; SMP 2a, 2b, 3e, 7a; PS 1b).

Procedure: The following is an example of how the game is played.

Teacher:	Java is 4 years old, and she has a birthday tomorrow. How old will Java be on her birthday tomorrow?
Jim:	5.
Teacher:	How do you know that?
Jim:	Because when we count, after 4 comes 5.
Teacher:	[*Presents a cupcake with four candles*] Last year Java had 4 candles on her cake like this. How many candles do we need to add to the cupcake to show Java is now 5 years old?
Jane:	1.
Teacher:	After Java turns 5, how old will Java be on her birthday next year?
Jane:	6 years old.
Teacher:	How many candles do we have to add to 5 candles to make 6 candles?
Jane:	1 again.
Teacher:	In yet another year how old will Java be?
Mary:	7.
Teacher:	Mary, how many candles do we have to add to 6 candles to make 7 candles?

Mary: 1 again.

Teacher: So, how many new candles did we have to add to Java's cupcake each year she had a birthday?

Students: We added 1 each year.

Activity 14. The "Confused Stair Builder"

Aim: Help children to discover via empirical induction that each successive step requires exactly one more block than its predecessor (the successor principle; SMP 2a, b, 3e, 7a; PS 1b).

Procedure: The following figures are similar to those from the "Confused Stair Builder" computer game (developed by the author for teaching experiments aimed at fostering the successor principle), and are used here to illustrate the steps for this activity. A teacher could implement the steps as a skit using stuffed animals, employing cardboard cutouts of the bear and monkey, or having children act as the bear and monkey. The objects in the figures could be represented by interlocking blocks, 8-by-8-inch interlocking jumbo blocks, toy blocks, stackable plastic milk crates, and so forth. Alternatively, a teacher could create a storybook by copying and pasting in the following figures and text.

Step 1. Set the stage for the successor question "What do I have to add to 3 to make 4?" (Part I).

Explain: "Bear wants to build a staircase so that he can reach the honey hive in the tree. Monkey wants to help but sometimes gets confused and makes mistakes. Can you help Bear and Monkey build the staircase to the honey hive?" Build a cardinality chart for 1 to 3 with (interlocking) blocks and the numeral cards 1, 2, and 3 (fig. 2.4).

Fig. 2.4. The cardinality chart

Step 2. Set the stage for the successor question (Part II).

Explain: "Bear says, 'I need to build the 4 step.' (Set the numeral 4 card immediately to the right of the numeral 3 card to indicate the next step to be built.) Monkey says, 'I know what to do; let me help.' And Monkey goes off to get blocks to build the 4 step" (fig. 2.5).

Fig. 2.5. Setting up to build the 4 step

Step 3. Set the stage for the successor question (Part III).

Explain: "Monkey brings back and stacks just 3 blocks for the 4 step. Bear says, 'Oh no, 3 blocks is not enough to make the 4 step'" (fig. 2.6).

Fig. 2.6. Monkey's incorrect 4 step

Step 4. Present the successor question "What do I have to add to 3 to make 4?"

Present additional successor questions to help children recognize that one is added to each number to create its successor. That is, continue to build up the staircase with a couple of additional steps. Encourage children to consider what is true for each example. It may help to include several hypothetical examples, such as: "If Bear wanted to build step 7, but Monkey brought him 6 blocks, how many more blocks would be needed to build the 7 step? How many would we have to add to a number to create the number that comes after it in the counting sequence?"

Informal Mathematical Induction and the Infinite Succession Principle

The successor principle is the conceptual basis for constructing a more precise mental representation of the counting sequence, namely viewing the sequence as the (positive) integer sequence n, $n + 1$, $(n + 1) + 1$, and so forth. This more precise representation of the counting numbers is necessary for the development of informal mathematical induction because it entails reasoning about the positive integers $(n, n + 1, [n + 1] + 1, \dots)$. Activity 15 illustrates how exploring big numbers and encouraging informal mathematical induction can be done together.

Activity 15. Exploring Large Numbers and Infinite Succession

Aim: Provide an opportunity to explore large numbers and—using the successor principle as a premise and children's knowledge of the rules for generating the counting sequence—engage in informal mathematical induction to devise the infinite succession concept (SMP 1b, 2a, 2d, 6a, 7a, 8b; PS 1a, 1b, 2a).

Procedure: With a small group of children, begin with warm-up questions such as "If we have two blocks and add one more, do we get a number that is larger or smaller than the one we started with? What is the larger number? If we have twelve blocks and add one more, do we get a larger or smaller number? What is the larger number? If we have twenty-two blocks and add one more, do we get a larger or smaller number? What is the larger number? If we have one hundred twenty-two blocks and add one more, do we get a larger or smaller number? What is the larger number?"

Teacher:	[*Continuing with the initial case*] What do you think is the largest number?
Aiden:	A hundred. [Note: Use whatever number the group considers the largest, even if it is a fictitious number such as a "gazillion."]
Teacher:	[*After the class generally agreed that the number 100 is the largest number*] Could we add one more to a hundred?

Madison:	I guess so.
Jaxon:	I agree. [*Others nod knowingly in agreement.*]
Teacher:	When we added one more to a hundred, do we get a larger or smaller number?
Class:	Bigger
Teacher:	What is the bigger number?
Madison:	[*After some murmuring*] A hundred and one.
Aiden:	We're adding, so it's bigger.
Teacher:	[*Introducing the next case*] Could we add one more to a hundred one?
Class:	[*Indicates agreement*]
Ava:	I think you can always add one more.

Alternatively, a teacher could say: "A really big number that comes after the hundreds, the thousands, the millions, and even the billions is called a *quintillion* (represented in the U.S. by the numeral 1 followed by 18 zeros). Do you think this is the last number—the largest number?" Proceed with the base step: "Could we add one more to [largest number suggested]? What if we added one more to [largest number suggested] do we get a bigger or smaller number? What is the bigger number?" If the children enjoy and are engaged by the process, follow up the base step with the inductive step once or twice more (with the same series of questions).

Another challenge would be to introduce *googol* (1 followed by 100 zeros) and then ask "Could we add one more to a googol ninety-nine? What number would we get?"

Possible follow-up probes (as needed): "Can we always add one to a really big number to make a bigger number? Why or why not? Is there a big number that we could not add one to to make it bigger? If we count and keep counting, will we ever get to the end of the numbers?" (If a student happens to respond to either of the last two questions with "infinity," tread carefully. *Infinity* commonly indicates "without any limit." In mathematics, it is treated as a number but not the same kind of number as a counting number.)

Instructional Implications of the Learning Progression

Fostering number sense and reasoning can and should go hand in hand. Clearly Nikki must have had a rich number sense to engage in informal mathematical induction and discover the concept of infinite succession. Many children come to school with a rich number sense, but many—perhaps most—do not. For these children, following the learning progression detailed in this section can help them achieve such a number sense (see the following case study of Barry). The examples in this section and the following case study make clear that instruction dealing with small numbers, counting, and numbers in general can provide children with many opportunities to engage in mathematical reasoning and can expand children's mathematical thinking as well as their mathematical knowledge. Engaging children in mathematical reasoning while teaching number and counting can help make such content instruction meaningful. Meaningful instruction is far more powerful than teaching counting and number content by rote, because children are more likely to remember the former and effectively apply it to new problems, situations, or tasks.

A learning progression can be invaluable in promoting number sense and reasoning. The case study of Barry, a primary-level student, underscores the value of using a learning progression to achieve a key instructional aim (comparisons of number neighbors such as six and seven) and to promote children's ability to reason about numerical relations. Barry was unable to make comparisons of more than with even small neighboring numbers, such as three and four. If he was placed on a staircase in which the steps were numbered or his tutor gave him a model (e.g., the cardinality chart shown below made up of interlocking cubes and numeral labels), he was able to choose the larger number. If he then worked on comparing neighboring numbers with the tutor, Barry gave the correct answer. However, after a week or so, he forgot all he had learned. Asked to count first and note the order sometimes helped him. Barry

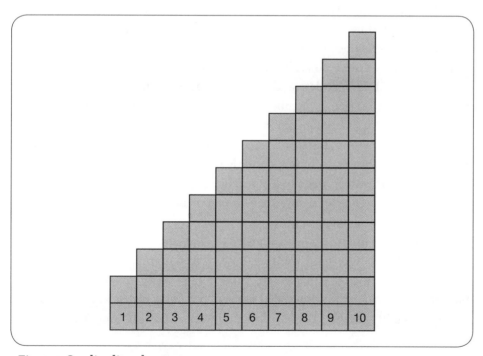

Fig. 2.7. Cardinality chart 1 to 10

could reliably count to twenty-one without error and sometimes to forty-one. He was capable of meaningful object counting (e.g., he could count out a specified number of objects fairly well) and could do simple addition on his fingers.

Why was Barry's instruction unsuccessful? Determining the larger of two neighboring numbers involves a logical reasoning process that is so automatic in adults that they may not appreciate the challenge it poses for young children. This is why a learning progression like that discussed previously can be valuable. It can help adults understand the complexity of the challenges facing children. Consider what is involved in answering the apparently simple question "Which is more, seven or six?" Two pieces of knowledge serve as the basis for fluently deducing the answer "Seven is more."

1. Knowledge of the increasing magnitude principle—an awareness that the counting sequence represents increasingly larger collections/quantities (e.g., seven is more than six because it comes after six in the counting sequence).

2. Knowledge of the counting sequence so fluent that it can be accessed at any number to determine the number after it—fluency with number-after relations (e.g., the number after six is seven).

Taken together these two pieces of knowledge allow children to logically deduce the larger of any two number neighbors.

A learning progression can be thought of as a staircase. Diagnosing a learning difficulty entails identifying what step on the staircase a child is on (has achieved) and then helping the child to take the next step on the staircase (achieve the next competence in the learning progression). If instruction is not well focused or at a level too far beyond the child's existing knowledge, it will be difficult for the child to understand and thus to acquire any long-term benefit from the instruction (e.g., remember and effectively apply the new information).

Consider Barry's situation and instruction, as summarized in table 2.1. Barry needed to focus on the second step in this table. Although he received some training at this step, it was apparently not sufficient to consolidate his achievement of this step. Much of his instruction was instead directed at step 4 even though he had not yet secured the developmental prerequisites for it; that is, he had not yet consolidated step 2 or step 3.

When Barry's instruction was adjusted to take into account the learning progression, he enjoyed step 2 training that involved snapping together blocks and comparing the lengths. Although it took about twenty sessions, he finally mastered the step. He then proceeded to make progress on step 3 and—at last report—was fairly reliable in determining number-after with random access.

Table 2.1. Barry's movement through the learning progression

Steps in the Learning Progression	Analysis of Barry's Attainment of Steps
Step 1. Meaningful object counting.	Barry appeared to have achieved this step.
Step 2. Increasing magnitude principle of counting and counting-based number comparisons.	Barry had not yet mastered this step. Some of his instruction involving the use of numbered steps and models targeted this step.
Step 3. Number-after Step 3(a) with a running start. Step 3(b) with random access.	It was unclear whether Barry had mastered number-after relations with random access or even with a running start.
Step 4. Mental comparisons of neighboring numbers.	The tutor's instruction initially and unhelpfully focused on this step.

Using Early Number and Arithmetic Instruction to Engage Children in Mathematical Reasoning

The theme of this third and final section is that regular content instruction provides many opportunities to involve children in mathematical reasoning of all types and that when it does, the meaningful learning of the content can greatly benefit. The first subsection illustrates how a key component of any learning progression—sweeping and important concepts ("big ideas")—can provide a basis for meaningful and inquiry-based instruction, including the often-problematic topic of fractions. The second subsection underscores how instruction based on a big idea and mathematical reasoning can help address another problematic area, namely memorizing the basic sums and differences. The third subsection illustrates how an often-overlooked form of logical reasoning can greatly enrich regular content instruction.

Number and Arithmetic Reasoning That Includes the Big Idea of Equal Partitioning

Big ideas include overarching concepts that provide the basis for understanding various concepts and procedures within a topic or even across topics. For example, the big idea of *equal partitioning* (subdividing a whole into equal-size parts) and its informal analogy of *fair sharing* can provide a meaningful basis for introducing such diverse ideas as division, fractions, and even numbers as well as many opportunities to engage in mathematical reasoning.

Informal Division

In prekindergarten–grade 2, children can be prepared to tackle formal (symbolic) division in a sensible manner if they are introduced to division informally as fair sharing. One way of doing this is to use children's literature as a source of sharing problems for students to solve. For example, instead of simply reading *The Door Bell Rang* to a class, a teacher can use the story line as a source of fair-sharing problems. When mother presents a brother and sister with a plate of twelve cookies, a teacher can pose the problem: How many cookies will each of the two children have if they share twelve cookies between them fairly? This problem actively involves children in thinking about how they can solve the problem, which is far more beneficial than simply having the answer read to them. Children as early as kindergarten can *intuitively* devise a dealing-out (divvying-up) strategy, and some can even logically justify the strategy (e.g., "If you give each person one each time, then everybody will have the same"). After sharing their strategies and solutions, and the class or group agrees that each child's share would be six, the teacher can read on, confirming their solution and then presenting the next problem: How many cookies will each of the three children (the two siblings and a newly arrived guest) get if they share twelve cookies among them fairly? The process can be repeated for sharing twelve cookies among six children and finally twelve cookies among twelve children.

Fractions

Fractions are particularly mystifying to many children for a variety of reasons. Relating fraction notation to the familiar situation of fair sharing can help students expand their understanding of what fractions can represent, make sense of fractions, reason about them more cogently, and avoid common problems.

Understanding division (quotient) and part-whole meanings of fractions. Children often narrowly view fractions in terms of a part-of-a-whole meaning and do not realize that these symbols can represent other meanings of rational numbers such as a division (a quotient meaning). Relating fractions to the familiar situation of fair sharing provides a basis for understanding that fractions can have a quotient as well as a part-whole meaning. Indeed, such problems can be a helpful transition from children's informal understanding of fair sharing to a formal understanding of a part-whole interpretation of fractions. Consider, for example, the following fraction problem: A pizza is shared fairly among three children. How much pizza does each child get? The fraction ⅓ can represent both the problem (the numerator 1 represents the a single pizza, the fraction bar means "shared fairly among," and the denominator shows the number of people sharing the pizza) and the solution to this fair-sharing problem (each child's share is one of the three equal pieces). Children can reason logically about even more challenging fair-sharing problems such as "Five children must share eight pizzas fairly. How much pizza will each child get?" (See Probe 9.3 on p. 9-10 of Baroody and Coslick [1998] for a discussion and depiction of three impressive informal solution strategies devised by second graders.)

Understanding the concept of equal parts. A common problem with fractions is illustrated by the case of Helene (Baroody 1987). Like many pupils, Helene did not realize that fractions describe a very special situation in which a whole is partitioned into *equal*-size parts. Overlooking this important constraint resulted in errors such as that illustrated in example A of exhibit 2.2.

Exhibit 2.2. The conceptual basis of part-whole fractions: equal partitioning (equal-size parts)

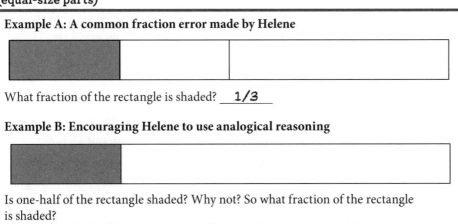

Example A: A common fraction error made by Helene

What fraction of the rectangle is shaded? __**1/3**__

Example B: Encouraging Helene to use analogical reasoning

Is one-half of the rectangle shaded? Why not? So what fraction of the rectangle is shaded?

In order to help the Helene understand the importance of equal partitioning, her tutor brought a candy bar to their session and offered to share it. Helene's joy quickly turned to dismay when the tutor cut off a fourth of the candy bar, handed it to the girl, and commented, "Here's your half." When Helene protested that she had been treated unfairly, the tutor asked, "What's unfair about it?" "The pieces have to be same size!" The tutor then engaged Helene in *analogical reasoning* to help her see the connection between her informal understanding of half based on fair sharing and representations of fractions she previously would have mislabeled (see example B of exhibit 2.2). Afterward, Helene exhibited a much better understanding of the idea that fractions involve a portion of a whole divided into equal-size parts.

Even and Odd Numbers

An inquiry-based approach to teaching even and odd numbers can provide opportunities to engage several different types of mathematical reasoning (see, e.g., Investigation 8.1 on p. 8-6 of Baroody and Coslick 1998).

Identifying even numbers via deductive reasoning. Even numbers can be defined as collections that can be shared fairly between two people. Odd numbers can be defined as collections that cannot be shared fairly between two people. Using these generalities as a premise, children can logically deduce for themselves whether a particular number up to twenty (or beyond) is even or odd by trying to fair-share that number of blocks with a partner.

Using informal mathematical induction to discover a general property about odd numbers. Mr. Sian asked his students to examine a number list up to 20 in which the odd numbers had been enclosed with red triangles and the even numbers with blue squares to see what they could find.

Marta: The first odd number, 1, is followed by the even number 2, and the next odd number, 3, is also followed by an even number 4. [*Using the two cases to notice a pattern in the counting sequence, she concluded*] So *any* odd number *must be* followed an even number.

Mr. Sian: [*Unclear about Marta's logic*] How do you know that it is true for all odd numbers?

Marta: [*Not having explicitly thought through the logic, delays*] Hmm . . .

Mr. Sian: [*Sensing that Marta is unsure how to begin*] Based on how we have defined *odd* and *even*, how do you know that all odd numbers are followed by an even number?

Marta: Oh, 1 [*drawing an o on the board*] is odd, because it can't be shared fairly between two people. One and one more [*drawing a second o on the board*] makes 2, an even, because it can be shared [*between two people*]. Two and one more [*illustrated by oo/o*] is 3, which is odd because it can't be shared fairly [*between two people*]. Three and one more [*illustrated by oo/oo*] is the even number 4. So if you keep on adding one more each time, you get an odd and then an even, another odd and then another even. Do I have to keep going?

Mental Arithmetic Reasoning That Includes the Big Idea of Solving Problems by Looking for Patterns

Many children have difficulty learning or remembering basic addition combinations such as 4 + 5 = 9 and 9 + 8 = 17 and related subtraction combinations such as 9 – 5 = 4 and 17 – 8 = 9. Trying to memorize the 121 basic sums to 20 and the 66 related differences is a formidable task that many children have little interest in tackling. Achieving fluency with basic combinations—another major goal of primary-level instruction—can and should go hand in hand with engaging children in mathematical reasoning. Specifically, the big idea is that mathematics is, at heart, problem solving, and, as we have seen, looking for patterns can be a useful strategy for doing so. Thus this strategy provides the rationale for the following heuristic, which applies to any domain of mathematics, including determining basic sums and differences: When solving mathematical problems, consider *looking for patterns*. Embodying numerous patterns and relations, the basic addition and subtraction combinations provide a rich basis for engaging pupils in inductive reasoning. Moreover, children use the arithmetic regularities they discover to intuitively invent logical reasoning strategies that enable them to use previously acquired knowledge to deduce unknown sums and differences.

Encouraging children to discover regularities and to use them to devise "thinking strategies" can make learning basic addition and subtraction a challenging and engaging detective and mind game instead of a drudgery. Not only is such an approach more motivating, but it is also more likely to result in sustained and growing fluency, that is, in retention of practiced combinations and the transfer of fluency to unpracticed ones. In brief, mental arithmetic instruction that focuses on looking for patterns (inductive reasoning) and using logical (deductive) reasoning strategies can provide children numerous opportunities to engage in mathematical reasoning *as well as* a meaningful and engaging basis for achieving fluency with basic sums and differences that is more effective than a drill-based approach.

The following are some general guidelines for teaching the basic addition and subtraction combinations:

1. *The goal should be fluency, not merely efficiency.* Fluency is the appropriate and adaptive—as well as efficient—use of knowledge. Appropriate use is the application of knowledge to relevant cases but not irrelevant ones. For example, appropriate use of commutativity (order does not affect the outcome) can be applied to addition (e.g., both 5 + 3 and 3 + 5 have the same sum) but not to subtraction (e.g., 5 − 3 and 3 − 5 do not have the same outcome). Adaptive use entails applying or adjusting knowledge to new cases or problems. For example, once children have inductively discovered commutativity, adaptive use implies that they can use this principle and knowledge of a known sum to deduce the sum of an unknown but related sum. For instance, if children understand that addition is commutative and know that the sum of 5 + 3 is 8, then they logically deduce the unknown sum of 3 + 5:

   i. Premise 1: If the sum of 5 + 3 is 8 and

   ii. Premise 2: the order of numbers does not affect the sum,

   iii. Conclusion: then the sum of 3 + 5 MUST be 8.

2. *Achieving fluency requires meaningful memorization, not memorization by rote.* If children learn basic combinations in terms of how they are related to each other and to other aspects of arithmetic understanding, then they will better be able to use combination knowledge in an appropriate and adaptive manner. In contrast, rote memorization may produce efficient recall (often only temporarily) but appropriate application only in familiar, not unfamiliar, contexts. For example, Torbeyns, Verschaffel, and Ghesquieere (2005) found that children who were taught the near-doubles sometimes used the strategy inaccurately.

Lukas: 7 + 8 is 13, because 7 + 7 is 14 and take away 1 is 13.

Teacher: Does everyone agree?

Wout: It's 15, because 8 + 8 − 1.

Teacher: You're saying you subtract 1 when you use the next larger double after the near double.

Daan: I got 15 too, but I thought 7 + 7 + 1.

Teacher: You're saying you add 1 when use the next smaller double before the near double.

Lukas: I'm confused; are you supposed to add or subtract?

In brief, arithmetic regularities and reasoning strategies should not be imposed on children. Instead, as much as possible, children should be encouraged to discover regularities or devise reasoning strategies themselves so that—unlike Lukas—they conceptually understand the procedure.

3. *The cornerstones of instruction geared to fostering meaningful memorization should be the following two big ideas for solving a new problem: (a) "look for patterns" (inductive reasoning) and (b) "consider how a new problem is like a previously solved problem and logically apply what you know" (deductive reasoning).* These problem-solving aids can be underscored by encouraging primary graders to share with the class any patterns or relations they induce (discover) or deduce (logically think out) when practicing arithmetic and then naming the discovered arithmetic regularity or strategy after the child or children or group that first brought it to the attention of the class. For example, if Azure recognizes that order does not matter when adding two numbers, the principle of additive commutativity can informally be called "Azure's pattern."

4. *Ensure that children have the developmental prerequisites necessary to discover regularities or deduce reasoning strategies.* For example, children need to achieve at least the first four steps in exhibit 2.1 on page 34 (i.e., up to and including fluently recognizing number-after relations) in order to discover or deduce even the most basic regularities.

5. *Create opportunities to discover arithmetic regularities by structuring practice to make patterns and relations more apparent.* For example, designing practice so that children practice a sum immediately after its commuted partner can facilitate discovery of additive commutativity. This is particularly important for less obvious patterns and relations that underlie larger sums and subtraction.

6. *Focus on process as well as outcome.* Although accurately determining sums and differences is critically important, perhaps more important is children's ability to justify an answer or strategy. Why does a particular answer or strategy make sense?

Following is a discussion of some common arithmetic regularities and reasoning strategies that are within the capabilities of primary children to discover or invent.

Adding 1

For children who have achieved at least step 4 (fluently specifying number-after relations) in the learning progression outlined in exhibit 2.1 (page 34), discovering the "number-after rule for adding 1" is accessible: The sum of adding 1 and another counting number n is the number after n in the counting sequence. This pattern is so apparent to developmentally ready children that they typically can use inductive reasoning to discover it with either guided or even unguided practice (Baroody, Eiland, Purpura and Reid 2012, 2013; Baroody, Purpura, Eiland and Reid, 2015). Once children discover the connection between adding 1 and their existing knowledge of number-after relations, they can use these two pieces of knowledge as premises to logically deduce the sum of any combinations involving the addition of 1 for which they know the counting sequence: The sum of adding 1 and another number n is the number after n (Premise 1); 158 + 1 involves adding 1 and another number (Premise 2); thus its sum *has to be* the number after 158—159! (Note that 159 is the necessary result of logically reasoning from the premises.)

Because children may overgeneralize a rule, using counterexamples is particularly important when learning a rule such as the "add 1" rule. For example, it is not uncommon for novices at mental addition to respond with the number after the larger addend for adding 0 combinations (e.g., 5 + 0 or 0 + 5 is 6), "add 2" items (e.g., 5 + 2 or 2 + 5 is 6), or other sums (e.g., 5 + 3 or 4 + 5 is 6). Therefore, it is important to include counterexamples to ensure that the "add 1" rule is not applied inappropriately.

Teacher:	What is your solution to the problem (involving 5 and 3 more)? Yes, Biyu.
Biyu:	6.
Teacher:	Does everyone agree? Yes, Chaitra.
Chaitra:	Disagree. It can't be 6, because 5 and 1 more is 6; so 5 and 3 has got to be more.

Doubles, Near-Doubles, and Make-Doubles

The doubles such as 5 + 5 and 6 + 6 are also relatively easy to learn because they can embody familiar real-world pairs of a set, such as five fingers on one hand and five on the other is ten fingers altogether and one row of six eggs and a second row of six eggs is twelve altogether (Rathmell 1978). The sums of doubles

1 + 1, 2 + 2, 3 + 3, . . . are the even numbers (and parallel the skip-count-by-two: 2, 4, 6, . . .), and the sum of a double is akin to the second count in various skip counts (e.g., 5 + 5 can be reinforced by knowing the skip-count-by-fives: five, ten).

Learning the doubles, however, is more difficult than learning the zero rule (adding 0 and a number leaves the number unchanged) or the number-after rule for adding 1, because the related knowledge they build on is less apparent or familiar to young children than that for the zero rule or the number-after rule. For example, young children may not readily recognize real-world embodiments of the doubles or the connection between the sums of doubles and the even number sequence. Moreover, they may not be highly familiar with skip counting by, for example, threes or fours. In brief, promoting fluency with the doubles will require highly guided instruction for most children.

Once children are fluent with the number-after rule for adding 1, the reverse of this rule (a number can be decomposed into the preceding number and 1), and the doubles, they can use these two pieces of knowledge as the premises for logically deducing the sum of any near-double, such as 5 + 6 and 7 + 6. The near-doubles reasoning strategy entails for 5 + 6, for instance, decomposing the 6 into 5 + 1, retrieving the sum of the double 5 + 5 = 10, and then adding the 10 and 1 to determine the sum of the near-double: 11. Note that if the child has not progressed to the point of fluency with the doubles, the reverse number-after rule for adding 1, and the number-after rule for adding 1, then it is less likely that he or she will discover or understand the near-doubles or be efficient in its application. This again underscores the importance of considering developmental readiness and the child's level on a learning progression.

Understanding the make-doubles reasoning strategy (e.g., 5 + 3 = [4 + 1] + 3 = 4 + [1 + 3] = 4 + 4 = 8), in effect, entails recognizing that when two numbers are two apart, taking 1 from one number and giving it to the other creates a double with the same total. Highly guided instruction can help primary-age children to discover this reasoning strategy. Children can be given a number of examples where the addends differ by two and asked to model them with blocks (e.g., "Dora and Nora found some candies left in a candy box. Dora took five candies and Nora took the three remaining candies."). Encourage the children to determine the total in each case (e.g., "How many candies did the girls find in all?"). Then ask them to consider whether or not the total changes of the larger collection shares a block with the smaller collection (e.g., "Dora realized that her share was unfair and so gave one candy to Nora."). After solving a number of similar problems, ask whether anyone noticed what happens when two collections that are 2 apart are shared fairly. With some additional exploration some children may discover that two numbers 4 apart can be transformed into a double by fairly redistributing 2, that two numbers 6 apart can be transformed into a double by fairly redistributing 3, that two numbers

8 apart can be transformed into a double by fairly redistributing 4, and that any two numbers that are an even number apart can be transformed into a double by fairly redistributing half their difference.

Subtraction as Addition

Basic differences are more difficult to learn than basic sums (Kraner 1980; Smith 1921; Woods, Resnick, and Groen 1975; see Cowan 2003 for a review). This is often the case even when subtraction is related to children's existing knowledge of addition in the form of the subtraction-as-addition strategy (e.g., for 8 − 5, think 5 and what make 8?) for two reasons. First, children view addition and subtraction informally as two different or independent actions or operations. Second, the relations between subtraction and addition are not obvious (Baroody 1999; Baroody, Ginsburg, and Waxman 1983; Canobi 2004, 2005, 2009; Henry and Brown 2008; Putnam, deBettencourt, and Leinhardt 1990).

Recognizing that addition and subtraction are interdependent operations and helping children achieve the Common Core standard 1.OA. B.4 (i.e., learn the addition-as-subtraction reasoning strategy) can be facilitated by guided instruction that highlights the following three relations:

1. *Complement principle and combination families.* The *addition-subtraction complement principle* can be illustrated by the proposition: if 3 + 5 = 8 or 5 + 3 = 8, then 8 − 5 = 3 or 8 − 3 = 5. This principle is the conceptual basis for the subtraction-as-addition reasoning strategy and implies that certain addition and subtraction combinations (e.g., 3 + 5 = 8, 5 + 3 = 8, 8 − 3 = 5, and 8 − 5 = 3) belong to the same "family."

2. *Empirical undoing and inverse concept.* Empirical undoing (e.g., adding 5 to 3 to make 8 and then taking 5 away from 8 to get 3 again) may be the basis for discovering the inverse concept (adding and then subtracting the same amount, or vice versa, leaves the original collection or number unchanged; for example, 3 + 5 − 5 = 3 and relating it to the complement principle (e.g., if 3 + 5 − 5 = 3, it makes sense that 3 + 5, which sums to 8, is related to 8 − 5 = 3; Baroody, Torbeyns, and Verschaffel 2005; Canobi 2004; Nunes, Bryant, Hallet, Bell, and Evans 2009).

3. *Part-part-whole relations.* For Piaget and others (Briars and Larkin 1984; Canobi 2005; Resnick 1983; Riley, Greeno, and Heller 1983), part-whole knowledge is the basis for understanding the complement principle, subtraction-as-addition strategy, and inversion. For example, if the part 3 and the part 5 make the whole 8, it makes sense that taking one of the parts, such as 5, from the whole leaves the other part 3. For this reason, curricula introduce "fact triangles" with the whole, such as 8, represented at the apex and each part, such as 5 and 3, at a base angle and, in the case

of *Everyday Mathematics* (University of Chicago School Mathematics Project 2005, see p. 503), the + and – symbols in the middle of the triangle, and the related equations (e.g., 5 + 3, 3 + 5, 8 – 3 = 5, and 8 – 5 = 3) to the side.

Using Qualitative Reasoning to Enrich and Promote Content Instruction

Qualitative reasoning is deductive reasoning that results in conclusions about arithmetical relations rather than a specific answer and includes reasoning about the direction of effect produced by operating on two (or more) numbers (e.g., whether an answer to five dollars take away two dollars must be smaller than, larger than, or equal to five). This type of logical reasoning can provide an important basis for introducing and assessing a variety of topics, including the meaningful learning of operation signs and estimation skills.

Introducing Symbolic Addition and Subtraction

Introducing symbolic or formal addition and subtraction—a major goal of primary-level instruction—can and should go hand in hand with engaging children in mathematical reasoning. Although children just beginning school may informally understand addition as adding more to a collection to make it larger and subtraction as taking away some from a collection to make it smaller, they may not know that the plus sign indicates addition or that the minus sign indicates subtraction. Activity 16 can serve to connect these formal symbols to children's informal understanding of arithmetic while engaging them in qualitative reasoning.

Activity 16. More Than or Fewer Than—Addition and Subtraction I

Aim: Help children assimilate the plus sign into their informal understanding of addition and the minus sign into their informal view of subtraction by engaging them in qualitative (logical) reasoning (SMP 1b, 2a; PS 1b, 2a, 3a).

Procedure: Present a series of written addition and subtraction expressions, such as 5 + 3, 5 – 3, 9 – 7, 9 + 7, 4 + 7, 13 – 5, one at a time. For 5 + 3 or 5 – 3, for instance, ask "Is the answer more than five or less than five ?" For 9 – 7 or 9 + 7, ask "Is answer more than nine or less than nine?" (Note that, initially, it would be helpful to contrast similar combinations, such as 5 + 3 and 5 – 3, by presenting them one after the other so as to clarify that the plus and minus signs indicate different operations.) In each case, ask for a justification: "Why?" A reasonable justification for why the answer to 5 + 3 is more than 5 would be as follows: "You started with five things [premise 1] and added more things to it [premise 2]. So now there has to be more than five things [logical conclusion]." A reasonable justification for why the answer to 5 – 3 is less than 5 would be the following:

"You started with five things [premise 1] and took some things away [premise 2], so now there must be fewer than five things left [deduction]." The activity can be played as game with several small groups (teams) or with individual children. Teams or players take turns answering, earning one point for a correct answer and a bonus point for a correct justification. An opponent can challenge an answer or justification and, if correct, earn a point or, if incorrect, lose a one.

Arithmetic Estimation

Many young children have difficulty estimating sums and differences because they believe only a single, correct or exact answer is desirable. This is often reinforced by guess-and-check estimation activities in which children make an estimate and then check it by determining *the* "correct answer." One way to counter the beliefs that undermine children's willingness to estimate is to encourage children to consider a *range of possible answers*. Activity 17 (a variation of activity 16) can provide students with experience in considering a range of answers to a challenging arithmetic problem by asking them to consider what the upper and lower limit might be and why. For $98 + 95$, for example, the sum could not be 200 or more (because $100 + 100 = 200$), the sum has to be more than 100 (because a lot is being added to 98, and 98 is almost 100 already) or—better yet—more than 180 (because $90 + 90$ is 180).

Activity 17. More Than/Fewer Than: Addition and Subtraction II

Aim: Help children assimilate the plus sign into their informal understanding of addition and the minus sign into their informal view of subtraction by engaging them in qualitative (logical) reasoning and using intuitive or even deductive reasoning to justify the range (SMP 2a, 1b; PS 1b, 2a, 3a).

Procedure: Explain, for example, "I will briefly show you an addition or subtraction expression with nine. Don't take the time to figure out the exact answer. Instead make a judgment *about* where the answer will be on this number list" (*point to Number List 1 [fig. 2.8]*). For instance, if the answer is "about 5, 6, 7, or 8," say "dark gray" and explain why 5 to 8 is a good estimate. Present a series of written addition and subtraction expressions with the same first addend, such as $9 + 3$, $9 - 3$, $9 - 7$, $9 + 7$, briefly one at a time. A good justification for $9 + 3$ is "Since you started with 9 items and only added a small number to them, the answer has to be a little past 9, so it's one of the light blue numbers." This format can be varied: a teacher could present a series of prompts such as "For $9 + 3$, are 5 to 8—the gray numbers—a good estimate or not, and why?" (e.g., "No, because they're smaller than 9, and the answer must be bigger than 9"). "Are 15 to 20—the dark blue numbers—a good guess or not, and why?" (e.g., "Kinda, but not really. Only a few were added to 9, and the dark blues are way more than 9").

Fig. 2.8. Number List 1

As children become familiar with the game, they can be presented more challenging approximations such as those required by, for instance, Number List 2 (fig. 2.9). Again, this game could be played either by having the pupil/ team/class choose a number range and justify the choice or by having the teacher choose a range and asking a pupil/team/class if it is a good estimate or not, and why. For the latter, a teacher could inquire "For 9 + 3, are 8 to 10—the white numbers—a good guess or not, and why?" (e.g., "No, because 8 is smaller, 9 is the starting amount, and you're not adding 0; and 10 is only 1 more than 9, not a few more"). "Is 20 or more a good estimate and why?" (e.g., "No, because 10 and 10 is 20, 9 is smaller than 10, and 3 is way smaller than 10").

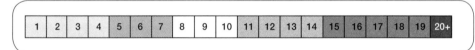

Fig. 2.9. Number List 2

Conclusion

Typically, number sense cannot be imposed on children by simply telling or showing them something about numbers. It usually stems from their making connections, discovering patterns, or discerning relations intuitively, inductively, deductively, or both inductively and deductively (mathematical induction). Fostering number sense, then, can and should go hand in hand with fostering mathematical reasoning. Number sense fosters mathematical reasoning, and mathematical reasoning, in turn, fosters number sense. Content instruction in this manner can also involve children in mathematical inquiry (problem solving, representing, and communicating as well as reasoning). This type of integrated instruction can help children to become more mathematically knowledgeable while becoming better mathematical thinkers.

Acknowledgments

Preparation of this manuscript was supported by a grant from the Institute of Education Science, U.S. Department of Education, through Grant R305A080479 ("Fostering Fluency with Basic Addition and Subtraction"). The opinions expressed are solely those of the authors and do not necessarily reflect the position, policy, or endorsement of the Institute of Education Science or the Department of Education.

References

Baroody, A. J. *Children's Mathematical Thinking: A Developmental Framework for Preschool, Primary, and Special Education Teachers.* New York: Teachers College Press, 1987.

Baroody, A. J. *A Guide to Teaching Mathematics in the Primary Grades.* Boston: Allyn & Bacon, 1989.

Baroody, A. J. "Children's Relational Knowledge of Addition and Subtraction." *Cognition & Instruction* 17 (1999): 137–175.

Baroody, A. J. "Discourse and Research on an Overlooked Aspect of Mathematical Reasoning." *The American Journal of Psychology* 118 (2005): 484–489.

Baroody, A. J., and R. T. Coslick. *Fostering Children's Mathematical Power. An Investigative Approach to K–8 Mathematics Instruction.* Hillsdale, N.J.: Erlbaum, 1998.

Baroody, A. J., M. D. Eiland, D. J. Purpura, and E. E. Reid. "Fostering Kindergarten Children's Number Sense." *Cognition and Instruction* 30 (2012): 435–470.

Baroody, A. J., M. D. Eiland, D. J. Purpura, and E. E. Reid. "Can Computer-Assisted Discovery Learning Foster First Graders' Fluency with the Most Basic Addition Combinations?" *American Educational Research Journal* 50 (2013): 533–573.

Baroody, A. J., H. P. Ginsburg, and B. Waxman. "Children's Use of Mathematical Structure." *Journal for Research in Mathematics Education* 14 (1983): 156–168.

Baroody, A. J., M.-L. Lai, and K. S. Mix. "The Development of Young Children's Number and Operation Sense and Its Implications for Early Childhood Education." In *Handbook of Research on the Education of Young Children*, edited by B. Spodek, and O. Saracho. Mahwah, N.J.: Erlbaum, 2006.

Baroody, A. J., D. J. Purpura, M. D. Eiland, and E. E. Reid. "The Impact of Highly and Minimally Guided Discovery Instruction on Promoting the Learning of Reasoning Strategies for Basic Add-1 and Doubles Combinations." *Early Childhood Research Quarterly* 30 (2015): 93–105.

Baroody, A. J., J. Torbeyns,and L. Verschaffel. "Young Children's Understanding and Application of Subtraction-Related Principles: Introduction." *Mathematics Thinking and Learning* 11 (2009): 2–9.

Briars, D. J., and J. G. Larkin. "An Integrated Model of Skills in Solving Elementary Word Problems." *Cognition and Instruction* 1 (1984): 245–296.

Bryant, P. E., and T. Trabasso. "Transitive Inferences and Memory in Young Children." *Nature* 232 (1971): 456–458.

Canobi, K. H. "Individual Differences in Children's Addition and Subtraction Knowledge." *Cognitive Development*, 19 (2004): 81–93.

Canobi, K. H. "Children's Profiles of Addition and Subtraction Understanding." *Journal of Experimental Child Psychology* 92 (2005): 220–246.

Canobi, K. H. "Conceptual-Procedural Interactions in Children's Addition and Subtraction." *Journal of Experimental Child Psychology* 102 (2009): 131–149.

Common Core State Standards (2011). "Common Core State Standards: Preparing America's Students for College and Career." Retrieved from http://www.corestandards.org/.

Cowan, R. "Does It All Add Up? Changes in Children's Knowledge of Addition Combinations, Strategies, and Principles." In *The Development of Arithmetic Concepts and Skills: Constructing Adaptive Expertise*, edited by A. J. Baroody and A. Dowker, pp. 35–74. Mahwah, N.J.: Erlbaum, 2003.

Donaldson, M. *Children's Minds.* New York: W. W. Norton, 1978.

Ennis, R. H. "Children's Ability to Handle Piaget's Propositional Logic: A Conceptual Critique." *Review of Educational Research* 45 (1969): 1–45.

Evans, D. W. "Understanding Zero and Infinity in the Early School Years." PhD diss., University of Pennsylvania (1983).

Evans, D. W., and R. Gelman. "Understanding Infinity: A Beginning Inquiry." Unpublished manuscript. Philadelphia: University of Pennsylvania, 1982.

Evans, J. St. B. T. *The Psychology of Deductive Reasoning.* Boston: Routledge & Kegan Paul, 1982.

Frye, D., M. Burchinal, S. M. Carver, N. C. Jordan, and J. McDowell. *Teaching Math to Young Children: A Practice Guide.* Washington, D.C.: National Center for Education Evaluation and Regional Assistance (NCEE), Institute of Education Sciences, U.S. Department of Education, 2013.

Hartnett, P., and R. Gelman. "Early Understanding of Numbers: Paths or Barriers to the Construction of New Understandings?" *Learning and Instruction* 8 (1998): 371–374.

Henry, V., and R. Brown. "First-Grade Basic Facts." *Journal for Research in Mathematics Education* 39 (2008): 153–183.

Inhelder, B., and J. Piaget. *The Growth of Logical Thinking.* New York: Norton, 1964.

Kaufman, E. L., M. W. Lord, T. W., Reese, and J. Volkmann. "The Discrimination of Visual Number." *American Journal of Psychology* 62 (1949): 953–966.

Kraner, R. E. "Math Deficits of Learning Disabled First Graders with Mathematics as a Primary and Secondary Disorder." *Focus on Learning Problems in Mathematics* 2, no. 3 (1980): 7–27.

Morris, A. K. "Development of Logical Reasoning: Children's Ability to Verbally Explain the Nature of the Distinction Between Logical and Nonlogical Forms of Argument." *Developmental Psychology* 36 (2000): 741–758.

Murata, A. "Paths to Learning Ten-Structured Understandings of Teen Sums: Addition Solution Methods of Japanese Grade 1 Students." *Cognition and Instruction* 22 (2004): 185–218.

Nunes, T., P. Bryant, D. Hallett, D. Bell, and D. Evans. "Teaching Children About the Inverse Relation Between Addition and Subtraction." *Mathematical Thinking and Learning* 11 (2009): 61–78.

Putnam, R. T., L. U. deBettencourt, and G. Leinhardt. "Understanding of Derived Fact Strategies in Addition and Subtraction." *Cognition and Instruction* 7 (1990): 245–285.

Resnick, L. B. "A Developmental Theory of Number Understanding." In *The Development of Mathematical Thinking,* edited by H. P. Ginsburg, pp. 109–151. New York: Academic Press, 1983.

Riley, M. S., J. G. Greeno, and J. I. Heller. "Development of Children's Problem Solving Ability in Arithmetic. In *The Development of Mathematical Thinking,* edited by H. P. Ginsburg, pp. 153–196. New York: Academic Press, 1983.

Sarnecka, B. W., and S. Carey. "How Counting Represents Number: What Children Must Learn and When They Learn It." *Cognition* 108 (2008): 662–674.

Schaeffer, B., V. H. Eggleston, and J. L. Scott. "Number Development in Young Children." *Cognitive Psychology* 6 (1974): 357–379.

Smith, J. H. "Arithmetic Combinations." *The Elementary School Journal* 10 (1921): 762–770.

Smith, L. *Reasoning by Mathematical Induction in Children's Arithmetic.* Oxford, U.K.: Pergamon, 2002.

Smith, L. "Children's Reasoning by Mathematical Induction: Normative Facts, Not Just Causal Facts." *International Journal of Educational Research* 39 (2003): 719–742.

Smith, L. "Mathematical Induction and Its Formation During Childhood." *Behavioral and Brain Sciences* 31 (2008): 669–670.

Sophian, C., and P. McCorgray. "Part-Whole Knowledge and Early Arithmetic Problem-Solving." *Cognition and Instruction* 12 (1994): 3–33.

Sophian, C., and K. I. Vong. "The Parts and Wholes of Arithmetic Story Problems: Developing Knowledge in the Preschool Years." *Cognition and Instruction* 13 (1995): 469–477.

University of Chicago School Mathematics Project. *Everyday Mathematics Teacher's Lesson Guide, vol. 1.* Columbus, Ohio: McGraw-Hill, 2005.

Woods, S. S., L. B. Resnick, and G. J. Groen. "An Experimental Test of Five Process Models for Subtraction." *Journal of Educational Psychology* 67 (1975): 17–21.

Wynroth, L. *Wynroth Math Program—The Natural Numbers Sequence.* Ithaca, N.Y.: Wynroth Math Program, 1986.

Algebraic Reasoning in Prekindergarten–Grade 2

Ana Stephens, with Maria Blanton

Many of us experienced algebra as an isolated grade 9 course that emphasized rote memorization, symbol manipulation, and artificial applications. It is now widely accepted that this traditional treatment of algebra must be replaced by an across-the-grades focus on *algebraic thinking* in which reasoning and sense making are front and center. There is evidence, furthermore, that students as young as prekindergarten can engage in and benefit from algebraic thinking. A focus on such thinking in the elementary grades serves two purposes. First, it helps prepare students for the more formal study of algebra that will come in their later studies of mathematics by building a fundamental understanding of core concepts. Second, many of the activities associated with algebraic thinking can actually deepen students' understandings of number and operation. Algebraic thinking should thus not be viewed as an "add-on" to an already full curriculum but rather as an opportunity to strengthen students' understanding of the existing curriculum and more.

What are these core algebraic concepts with which young elementary students can engage that will both lay the foundation for future study and strengthen their existing understanding of arithmetic? In our work developing an "Early Algebra Learning Progression" (Fonger et al. 2015), we have identified four "big ideas" that we believe are important and appropriate for students in prekindergarten–grade 2 to develop:

- *Equality, expressions, equations, and inequalities.* Students develop a deep understanding of the meaning of the equals sign, write and interpret arithmetic and algebraic expressions, and come to recognize the structure underlying arithmetic and algebraic equations and inequalities.

- *Generalized arithmetic.* Students identify and express arithmetic relationships, including fundamental properties of number and operation.

- *Functional thinking.* Students identify relationships between covarying quantities and express these relationships using a variety of representations.
- *Variables.* Students express mathematical relationships in succinct ways using symbolic notation.

The foregoing big ideas can be found throughout the algebra standards in the *Common Core State Standards for Mathematics* (CCSSM; NGA Center and CCSSO 2010). While they will each be discussed in turn in this chapter, they are not mutually exclusive and are actually quite difficult to separate. For example, students' work in the contexts of both generalized arithmetic and functional thinking often leads them to write expressions and equations and to express their generalizations using variables.

Key elements of reasoning and sense making evidenced and needed by students engaging in algebraic thinking around these four big ideas in the elementary grades include the following:

- *Generalizing mathematical relationships and structure.* Students notice relationships or patterns in arithmetic operations, expressions, equations, or function data that can be generalized beyond the given cases.
- *Representing generalizations.* Students use various notational systems—words, symbols, tables, graphs, and pictures—to represent their generalizations.
- *Justifying generalizations.* Students justify their generalizations and move toward making general arguments that go beyond examples-based reasoning.
- *Reasoning with generalizations.* Students make use of their generalizations to solve problems.

These elements of reasoning and sense making are very closely linked to the Common Core State Standards for Mathematics' Standards for Mathematical Practice (SMP) and the National Council of Teachers of Mathematics' (NCTM) Process Standards (PS; 2000). For example, when generalizing and representing relationships, students are "looking for and expressing regularity in repeated reasoning" (SMP 8) using a variety of representations (PS 5); when justifying generalizations, students are "constructing viable arguments" (SMP 3) and engaging in reasoning and proof (PS 2); and when students are reasoning with generalizations they are often "looking for and making use of structure" (SMP 7) in a way that aids problem solving (SMP 1; PS 1).

This chapter is organized around four "big ideas": (1) equality, expressions, equations, and inequalities; (2) generalized arithmetic; (3) functional thinking; and (4) variables. Within each of these sections, examples of mathematical tasks, student work, and classroom dialogue illustrating key elements of reasoning and sense making are discussed with explicit links to the Standards for Mathematical Practice and the Process Standards. This chapter will thus describe what reasoning and sense making "look like" in the domain of algebraic thinking in prekindergarten–grade 2 and highlight opportunities to engage students in this thinking.

Equality, Expressions, Equations, and Inequalities

Expressions, equations, and inequalities are necessary tools that enable students to express, justify, and reason with generalizations. Expressions, equations, and inequalities play a role in developing students' understandings within the other big ideas of generalized arithmetic, functional thinking, and variables and will be discussed within each of those sections. The focus here will be on the role that expressions, equations, and inequalities can play in helping students make sense of mathematical equality, specifically the meaning of the equals sign.

Developing an Understanding of the Equals Sign

One of the Standards for Mathematical Practice states that students must "attend to precision" (SMP 6), which includes "using the equals sign consistently and appropriately" (SMP 6c). A solid understanding of the equals sign and its role in a variety of mathematical equations is critical both to developing a fundamental understanding of arithmetic and to learning algebra. It is an important tool in expressing and reasoning with generalizations. Unfortunately, many elementary students—and even older students—harbor serious misconceptions about the meaning of the equals sign. Carpenter, Franke, and Levi (2003) found, for example, that when asked what number should be put in the box to make the number sentence $8 + 4 = \square + 5$ true, the vast majority of first through sixth graders gave responses of 12 (the sum of 8 and 4), 17 (the sum of all three given numbers), or 12 *and* 17. These students are said to have an *operational* view of the equals sign because they view the equals sign as a stimulus to "do something" (i.e., compute) rather than a *relational* view, which implies an understanding of the symbol as denoting the relation between two equal quantities. This misconception both hinders students' abilities to learn and express arithmetic ideas flexibly and causes difficulties when students begin a more formal study of algebra.

Researchers have found that students who have an operational view of the equals sign experience more difficulty solving algebraic equations when they reach middle school (Knuth et al. 2006). The good news for teachers of young students is that the operational view is often less entrenched than is the case with older students. In fact, in one study, first graders were found to be more successful than second- and third-grade students in coming to view the equals sign as a symbol denoting an equivalence relation rather than a signal to compute because they had less experience with the symbol and therefore had not yet developed deep-seated misconceptions (Baroody and Ginsburg 1983).

What is the source of the operational view of the equals sign, and what can be done to help students to reason consistently about the symbol's meaning? First, consider the types of experiences students typically have with this symbol. Students who are asked to complete worksheets of "fill-in-the-blank" number sentences such as $1 + 3 = $ ___, $5 + 2 = $ ___, and $9 - 4 = $ ___ are only seeing the equals sign in one context. It is not surprising that students exposed exclusively to this type of problem will come to view the equals sign as a signal to compute or "put the answer." In this limited context, such a conception actually serves them well.

It is critical, then, to ensure that students have exposure to the equals sign in a variety of equation contexts from the very beginning of their experiences with the symbol. True/false and open number sentences are particularly good vehicles for engaging students in a discussion of the meaning of the equals sign and challenging any existing operational conceptions. Consider the following tasks and how students might make sense of them:

True/False Number Sentences	Open Number Sentences
$3 + 5 = 8$	$6 + 4 = \square$
$3 + 5 = 0 + 8$	$6 + 4 = \square + 3$
$8 = 3 + 5$	$\square = 9 + 3$
$8 = 8$	$5 + 3 = \square + 3$

Even students with an operational view of the equals sign will be able to state that $3 + 5 = 8$ is true or that a 10 should be put in the box in $6 + 4 = \square$. How would they respond to the less familiar number sentences? Think about the rich discussions that might ensue as students consider the following:

- If $3 + 5 = 8$ is true, what about $3 + 5 = 0 + 8$? Is this number sentence false because $3 + 5$ is not equal to 0, or is it true because $0 + 8$ is the same as 8?

- Should 10 be put in the box in $6 + 4 = \square + 3$ because $6 + 4$ is equal to 10? But what about the 3? What is its role in the equation?

- Is $8 = 3 + 5$ true, or is it "backward"? If $3 + 5 = 8$ is true, isn't $8 = 3 + 5$ true as well?

- What can we put in the box in $\square = 9 + 3$? Can we write number sentences this way?

- Is $8 = 8$ true even though there is nothing to "do"?

- Should we put an 8 in the box in $5 + 3 = \square + 3$, since $5 + 3 = 8$? But, wait. . . . Isn't $5 + 3$ the same as $5 + 3$?

Another approach to developing students' understanding of the equals sign—as well as symbols indicating inequalities—involves engaging students in thinking about quantities and making comparisons. First-grade students participating in the *Measure Up* (Dougherty 2008) project learn to compare and describe measurements in terms of "equal to," "not equal to," "greater than," and "less than" (SMP 2a). The quantities described are unspecified (e.g., students might compare two jars holding different amounts of liquid); thus, students are guided to name the unknown quantities with letters (PS 5a) and express their relationships symbolically (SMP 2b, c; PS 5a). The relationship between two unknown volumes, A and B, can be expressed as $A = B$, $A \neq B$, $A < B$, or $A > B$. The use of the equals sign in this context helps students to view the symbol as expressing a relationship between two quantities as opposed to a symbol indicating that an operation needs to be performed. Other researchers (Hattikudur and Alibali 2010) found that third- and fourth-grade students developed a deeper conceptual understanding of the equals sign when they learned about this symbol alongside the inequality symbols.

In addition to posing true/false and open number sentences specifically designed to engage students in thinking about the equals sign, and comparing this symbol to those signaling inequalities, there are opportunities in existing curricula and everyday instruction to help students develop a relational understanding of the equals sign and challenge their misconceptions. Consider, for example, the equation "strings" students often write to keep track of a series of calculations. For example, suppose four students have 5, 3, 6, and 10 pennies, and we want to determine how many pennies they have altogether. A student might represent his or her calculations in response to this problem in the following way:

$$5 + 3 = 8 + 6 = 14 + 10 = 24$$

This student's recording provides an opportunity to consider whether the representation makes sense and to discuss the meaning of the equals sign. In fact, what this student produced is not mathematically correct, as it implies that $5 + 3 = 24$. Multiple calculations cannot be represented in a single equation; a series of equations is needed:

$$5 + 3 = 8$$
$$8 + 6 = 14$$
$$14 + 10 = 24$$

Such equation strings can be used as opportunities to help students attend to precision and use the equals sign consistently and appropriately.

Building Equations from Equivalent Expressions

Next, consider an activity common to many first-grade classrooms: helping students develop fluency in adding numbers that sum to ten. While some teachers might approach this task by providing students a list of computations to perform, the potential exists to engage students in examining expressions, equations, and their conceptions of the equals sign.

Suppose that, rather than having students compute sums, you ask them to generate additive expressions equivalent to 10 (e.g., 3 + 7, 8 + 2). This might be presented in a familiar context such as in the following task posed to a first-grade class (Russell, Schifter, and Bastable 2011a, pp. 5–6):

> You have 10 vegetables on your plate. Some are peas and some are carrots. How many peas and how many carrots could you have?

Students first settled on 5 peas and 5 carrots as "the" answer. The teacher explained, "I got blank faces when I asked if there was another way to solve the problem. It seemed as if these children were used to getting one answer and moving on . . ." (Russell et al. 2011a, p. 6). After prompting students to think of more possibilities, more solutions were generated, including 9 peas and 1 carrot and 3 peas and 7 carrots.

This activity can be extended to help students strengthen their understanding of the equals sign. From the expressions representing the total number of peas and carrots (e.g., 5 + 5, 9 + 1), equations with operations on both sides of the equals sign (e.g., 5 + 5 = 9 + 1) can be generated and discussed. A student with an operational view of the equals sign might say that 5 + 5 = 9 + 1 is false because 5 + 5 does not equal 9. However, the context of "making 10" might provide the support such a student needs to confront his or her misconception. If students agree that 5 + 5 and 9 + 1 both sum to 10, they might be more apt to accept the equation as true. Some teachers use the following notation to support students' thinking:

Notice the role that expressions and equations play in this example. Rather than completing a series of computations in which a single numerical answer is expected, students are asked to express ten in multiple ways. If students regularly use expressions to represent values in this way, they will be better prepared to use algebraic formulas (Smith and Thompson 2008). The equivalent expressions are then the building blocks that allow students to construct equations and engage with the meaning of the equals sign. Such an activity addresses the need to develop students' fact fluency while also laying the foundation for important algebraic work with expressions and equations. As Smith and Thompson (2008) argue, "We cheat our students if our problems are always requests for calculations" (p. 114). Indeed, as will be illustrated throughout this chapter, sometimes it makes sense *not* to compute.

Examining Equations in Their Entirety

Once students have developed a relational understanding of the equals sign, their work with equations is just beginning. Understanding the equals sign as a symbol denoting the relationship between two equal quantities enables students to look across the equals sign and consider equations in their entirety as mathematical objects (SMP 7c, d). Refer back to the open number sentence $8 + 4 = \Box + 5$. We know that students with an operational view of the equals sign are not apt to place the correct number, 7, in the box. What about students who have a relational view of the equals sign and do place 7 in the box? How might they be thinking about the problem? Carpenter et al. (2003) share the explanations of two students, Ricardo and Gina:

Ricardo: 8 and 4 is 12. So I had to figure out what to go with 5 to make 12, and I figured out that had to be 7.

Gina: I saw that the 5 over here [*pointing to the 5 in the number sentence*] was one more than the 4 over here [*pointing to the 4 in the number sentence*], so the number in the box had to be one less than the 8. So it's 7 (p. 13).

While Ricardo and Gina both demonstrate a relational understanding of the equals sign (i.e., they understand the symbol as expressing an equivalence relation between the two sides of the equation), Gina shows more flexibility than Ricardo. She understands the equals sign as a symbol expressing an equivalence relation between two expressions rather than two calculations. She considers $8 + 4$ as more than a calculation to be carried out and can reason with this uncalculated expression. This is not to say that Ricardo could not solve the problem as Gina did. The calculation in this example is fairly straightforward. It is easier to observe whether children actually need to calculate when larger numbers are used (Carpenter et al. 2003).

Consider the following true/false and open number sentences and their potential to further students' understandings of equations:

True/False Number Sentences	Open Number Sentences
$10 + 11 = 9 + 12$	$8 + 15 = \Box + 16$
$78 + 20 = 79 + 19$	$53 + 23 = 52 + \Box$
$43 + 66 = 44 + 65$	$\Box + 99 = 56 + 100$

While an open number sentence like $8 + 4 = \Box + 5$ can be solved fairly easily with computation, the examples above would be more cumbersome for young students to compute. The relationships between these larger numbers, however, encourage students to study the equations in their entirety and develop more efficient strategies. For example, given $10 + 11 = 9 + 12$, students might initially find the sum on each side to determine if the number sentence is true or false. On further reflection, however, students might look across the equals sign and notice a relationship: 10 is one more than 9, and 11 is one less than 12, so the number sentence is true. Number sentences with even larger numbers, such as $43 + 66 = 44 + 65$ and $53 + 23 = 52 + \square$ can push students to find more efficient methods by examining relationships across the equals sign and considering the number sentence as a whole object (SMP 7c, d).

Stepping back to examine an equation in its entirety and consider the relationships present will be worthwhile for students both in elementary school and in their later experiences with more formal algebraic equation solving. Consider an equation such as $12(x - 8) = 2(x - 8) + 20$ and two possible solution strategies:

Solution 1	Solution 2
$12(x - 8) = 2(x - 8) + 20$	$12(x - 8) = 2(x - 8) + 20$
$12x - 96 = 2x - 16 + 20$	$10(x - 8) = 20$
$12x - 96 = 2x + 4$	$x - 8 = 2$
$10x - 96 = 4$	$x = 10$
$10x = 100$	
$x = 10$	

The first few steps of the first approach are in some ways similar to Ricardo's approach. These steps involve considering each side of the equation separately and simplifying each side as much as possible before considering any relationships across the equals sign. The second approach resembles Gina's, in that "whole-equation" thinking is employed at the start. Taking advantage of mathematical relationships across the equals sign allows the equation to be solved much more efficiently.

As we have seen, expressions and the equals sign combine to form equations that serve as important objects for students to reflect upon in the context of their study of arithmetic and in preparation for a more formal study of algebra. Equality, expressions, equations, and inequalities are critical components of students' work with the remaining big ideas to be discussed in this chapter.

Generalized Arithmetic

Students spend much of their time in mathematics classes performing individual computations and producing "answers." An important part of algebraic thinking, however, involves looking across multiple examples and computations, noticing patterns and underlying structure (SMP 7a), articulating and representing generalizations about the mathematical relationships observed (PS 3a, c; PS 5a), and justifying those generalizations (SMP 3a, b, d; PS 2a, b, c). These are activities with which even very young students can engage.

Engaging with the Fundamental Properties

Consider an example in which a common mathematical task requiring a single numerical answer can—with accompanying classroom discussion—lead students to engage in multiple aspects of algebraic reasoning and sense making. Students enter school with informal mathematical knowledge derived from experience. They can join, separate, and compare quantities by counting. A teacher of prekindergarten or kindergarten students might help students employ this knowledge by posing a question such as the following:

> *If Evelyn had 2 books and borrowed 6 more books from the library, how many books does she have now?*

There are a variety of strategies and tools that students might use to solve this problem. Building on their informal knowledge, many young students might model the action in the story by counting out two "books"—represented by manipulatives, drawings, tally marks, or fingers—adding six "books" to this set, and then counting the total number of books. Others might add on from the first number given and think "2, [*pause*] 3, 4, 5, 6, 7, 8." Finally, some students might choose the more efficient strategy of adding on from the larger number. A student using this strategy might think "6, [*pause*] 7, 8." This collection of strategies offers an opportunity to discuss a fundamental property of number and operation.

In response to a student who adds on from the larger number, a teacher might ask "Can we do that?" The student might suggest that when you are adding two numbers, the order in which you add doesn't matter and that you will get the same answer either way. This fundamental property is known as the commutative property of addition.[1] While the name of the property is not necessarily important for students to recite, what is important is the opportunity the property provides to engage students in reasoning and sense making. Students who use the efficient "counting-on-from-larger" strategy and state their reasoning explicitly like this are engaging in algebraic reasoning by generalizing a mathematical relationship and reasoning with a generalization, that is, making use of structure to solve a problem (SMP 7a; PS 1a). A classroom in which this kind of discussion is taking place both prepares students for more formal algebraic reasoning where noticing structure and commonalities across cases is the norm and strengthens their arithmetic skills by exposing them to an efficient addition strategy and effectively cutting in half the number of "number facts" they need to learn.

It might be tempting to introduce students to the commutative property of addition and teach the counting-on-from-larger strategy directly. It is reasonable to want to help students compute efficiently, and it might feel as though waiting for them to "discover" the property and the efficient strategy on their own will take too long. Such strategies, however, make much more sense to students and serve as a more effective conceptual foundation for future learning, when they arise from students' own thinking about a problem (Carpenter et al. 1999). Arithmetic tasks should be chosen or designed in such a way that, through rich classroom conversations, children have the opportunity to notice the fundamental properties and think about efficient strategies for computing. We know that even very young students, when provided well-designed tasks supported by classroom instruction, are capable of recognizing structure (SMP 7a), making generalizations, and generating meaningful and efficient strategies on their own (SMP 1c, h; PS 1a). Even if only one student proposes a particular strategy, the opportunity arises to engage the rest of the class in thinking about it and building new mathematical knowledge (PS 1b).

There are several other fundamental properties of number and operation, or generalizations derived from those properties, accessible to children in kindergarten–grade 2. For example, consider the following generalizations generated by students in a combination first- and second-grade class (Carpenter and Levi 2000):

- Zero plus a number equals that number.
- If you subtract the same number from the same number [i.e., a number from itself], you will get zero.
- If you subtract zero from a number, you will end up with the same number.

These generalizations and others can be elicited through carefully chosen word problems like the foregoing commutative property of addition example or through the use of true/false and open number sentences. For example, a teacher might pose the seemingly "easy" number sentences $4 + 0 = 4$, $12 + 0 = 12$ and $56 + 0 = 56$ and ask students if they are true or false. Seeing several examples gives students the opportunity to generalize that "when you add zero to a number, you get the number you started with" (SMP 7a). Notice the importance of task selection in these examples. A teacher who wishes to "set up" her students for a conversation about the commutative property of addition, for example, would not ask "If Evelyn had six books and borrowed two more books from the library, how many books does she have now?" (i.e., the same task as that previously discussed but with the numbers reversed) because adding on from the larger number does not require the use of this property.

We have seen how children can both generalize mathematical relationships and structure and reason with their generalizations in solving problems

(SMP 7a). Next, let's extend the commutative property of addition example by considering how students might go about justifying their generalizations (SMP 3a, b, d; PS 2a, b, c) and what questions might encourage them to do so.

Once students have decided that "If Evelyn had 2 books and borrowed 6 more books from the library, how many books does she have now?" can be solved by counting on from the larger number and have generalized beyond this one example by stating that the order in which you add two numbers does not matter, a teacher might be satisfied that they have successfully adopted an efficient problem-solving strategy. However, this problem-solving and generalization exercise can be extended to engage students in two more important aspects of algebraic reasoning and sense making: justifying the generalization (SMP 3a, b, d; PS 2a, b, c) and representing the generalization using variables (SMP 2b, c; PS 3a, c; PS 5a).

First, students might be asked questions about their generalizations that extend beyond "Can we do that?" to *Why* can we do that?" and "Will that *always* work?" These are sometimes difficult questions for children because they do not necessarily have a way of talking about numbers in general and need to rely on specific examples (Carpenter et al. 2003). Sometimes, however, students' justifications will not depend on the specific numbers used in a given example. For example, consider how Alicia justified the generalization that naturally arises from the word problem we have been discussing (from Carpenter et al. 2003, p. 89):

> When you add two numbers, you can change the order of the numbers you add, and you will get the same number.

Alicia: It's like this. If you have 7 plus 5 [*puts out a set of 7 blocks and to the right of it a set of 5 blocks*], look you can move them like this [*moves the set of 5 blocks so that they are to the left of the set of 7 blocks*]. Now it's 5 plus 7, but it's still the same blocks. It's going to be the same when you count them all. It doesn't matter which you count first; it's still going to be the same.

Ms. P.: Okay, I see how that works for 7 and 5, but how do you know that is true for all numbers?

Alicia: It doesn't matter how many are in the groups. It could be any number. You are just moving them around like I did there; they are still the same blocks no matter what number you use.

Notice that while Alicia used seven blocks and five blocks in her justification, there is nothing special about these specific numbers. She could have chosen any numbers, and she clearly understands this. Furthermore, she did not need to calculate in the process of justifying her generalization. This type of justification is sometimes referred to as "representation-based" reasoning (Russell,

Schifter, and Bastable 2011b), because it relies on the use of a physical or visual representation as a bridge to a general argument.

Another algebraic reasoning activity connected to the fundamental properties with which students can engage is representing their verbal generalizations using symbols (SMP 2b, c; PS 3a, c; PS 5a). One way to begin this process is to start with a specific generalization that students have already made and ask them to think of an open number sentence that says the same thing. The zero property of addition is a good place to start, as it is a fairly simple generalization that only involves one variable. Consider the following discussion in a combination first- and second-grade class about whether "Zero plus a number equals that number" and "A number plus zero equals that number" are the same or different generalizations (Carpenter and Levi 2000, p. 13):

Laura:	They are different. The first one is like zero plus number equals number [*she writes* $0 + \# = \#$ *on the chalkboard*], and the other is like number plus zero equals number [*she writes* $\# + 0 = \#$].
Ms. Keith:	Can anyone else write those [generalizations] with something other than a number sign?
	[Mitch wrote $* + 0 = *$ and $0 + * = *$. Carla wrote $s + 0 = s$ and $0 + s = s$, and Jack wrote $m + 0 = m$ and $0 + m = m$.]
Ms. Keith:	Jack just wrote zero plus m equals m. What were you using the m to mean?
Jack:	I meant it to mean any number.
Ms. Keith:	Then we could put any number in there? What if I put a 2 here and an 8 here?
Laura:	No . . . you could put a 2 here and a 2 here. You have to put the same thing in both places.

The symbolic representations provide students with a way to be precise about their generalizations and to express them in a succinct manner. As they get older, for example, it will be much easier to express the distributive property as $a(b + c) = ab + ac$ than to try to express the idea in words. The use of variables in expressing generalizations will be revisited in the last section of this chapter.

Examining Relationships Among Operations

Opportunities to look for mathematical structure and generalize mathematical relationships extend well beyond students' work with the fundamental properties. Prekindergarten–grade 2 students are often asked to combine or separate sets of objects, but they are not always asked to consider the relationship between addition and subtraction. Asking them to do so, however, provides a context for engaging students in algebraic reasoning and sense making as well as for strengthening their knowledge of arithmetic. See chapter 2 in this book or Schifter (1999) for examples of students engaged in this kind of work.

Working with Inequalities

In addition to working with equations, algebra students are also expected to work with inequalities. As we have seen, this work can begin at a very young age by asking students to do something that comes naturally: comparing quantities (SMP 2a). Consider the game "double compare," in which two students hold piles of cards bearing numerals from 1 to 6 and a picture of that number of objects. Players lay down the top two cards from their piles, and the player with the higher total when the numbers on the cards are combined says "me." Note what happened when a class of kindergarten students played the game:

> No sooner was the game in question under way when [the teacher] realized that there were several pairs of students saying "me" or "you" before they could possibly have had time to find the sum of their numbers—but they were always right! For example, when Martina had 6 and 2 and Karen had 6 and 1, Karen quickly said, "You." When . . . asked how she knew, Karen pointed to the 2 and said, "This is big. Even though these are the same [the sixes], this [the 6 and 2] must be more." (Schifter et al. 2008, pp. 264–65)

While the original intention behind this task was to give kindergarteners an opportunity to practice counting up and finding totals, the teacher noticed that her students were sometimes able to successfully play the game without doing so. They were instead able to engage in algebraic reasoning by implicitly working with the generalization that if two of the numbers are the same, they can "ignore" them and compare just the other numbers. They were able to make this generalization explicit when the teacher asked, "Does this only work for six?"

More formally stated, the students were working with the generalization that if one number is greater than another and if the same number is added to each, the first total will be greater than the second. This statement is true for any three numbers and can be expressed as the following: *For any numbers, a, b, and c, if $a > b$, then $a + c > b + c$*. While kindergarteners would not be asked to state the generalization in this way, providing students opportunities to work with such regularities and examine structure in the number system will ease the transition to more formal approaches in their future work.

Examining Relationships Between Even and Odd Numbers

Another opportunity for students to look for mathematical structure and generalize mathematical relationships that presents itself in the context of elementary-school mathematics is the examination of even and odd numbers.

Consider the scenario posed by one first-grade teacher (Bastable and Schifter 2008) who painted pictures of snowmen on dried lima beans and told her students that the snowmen had received an invitation to attend the Snow Ball but only if they came in partners. This scenario fostered students' engagement in several aspects of algebraic thinking:

> This problem led to several days of thinking about odd and even numbers. They began to make observations about what numbers of snowmen could and couldn't go to the ball. [Using their lima bean snowmen,] soon a few children began exploring why six [snowmen] could [go to the ball] but not seven; four but not five; and came up with a rule: "each time you add one number to a group that can go, you get a group that can't." (p. 176)

While not yet using the terms *even* and *odd*, students were making generalizations about classes of numbers. The rule previously described could be restated as: "An even number plus one equals an odd number." Another student noticed that if you add two to a number of snowmen that can go, you get another number that can go (i.e., an even number plus two equals an even number), while another student used the lima beans to show that if you add together two groups that can't go, you get a group that can go (i.e., an odd number plus an odd number equals an even number).

As is the case with generalizations involving the fundamental properties, it can be difficult for students to justify their generalizations about even numbers and odd numbers using general arguments rather than specific examples. Consider Kimberly, a second-grade student, and her teacher having the following conversation to prove the generalization "When you add two even numbers, you get an even number" (Carpenter et al. 2003, pp. 87–88):

Ms. V.:	How would we show that is always true no matter which even numbers we use?
Kimberly:	We could try it.
Ms. V.:	What do you mean we could try it? We could add two even numbers and that would show that it was always true?
Kimberly:	No. We could all try it. Everyone in the class could do some. We could each try a lot of numbers.
Ms. V.:	So if we have a lot of numbers, will that show it is always true?
Kimberly:	Well, if all the answers were even it would.
Ms. V.:	Could you try all the numbers?
Kimberly:	No. But you could try a lot of numbers.

Ms. V.:	How about the numbers we didn't try? How could we be sure it worked for them?	
Kimberly:	Well if it works for all the numbers we try, it probably works for the other numbers, too.	
Ms. V.:	Suppose someone in the class said there might be an even number that we haven't tried that it won't work for?	
Kimberly:	I'd tell them to show me the number.	

Kimberly is engaged in justification by example and seems to believe that if no one can find an example that does not fit the generalization, the generalization must be true. This is another example of a problem in which specific numbers can be used in a general way to make a general argument. If students think of even numbers as consisting of groups of two (e.g., partnered snowmen) and odd numbers as consisting of groups of two with one "left over" (e.g., a snowman without a partner), they can form various combinations of even and odd numbers (represented by drawings, interlocking cubes, or lima beans, for example) without being concerned about the specific numbers. A student might think in the manner illustrated in the chart below:

Group 1	Group 2	Group 1 + Group 2
Snowmen in partners (even)	Snowmen in partners (even)	Snowmen in partners (even)
Snowmen in partners (even)	Snowmen in partners with one left over (odd)	Snowmen in partners with one left over (odd)
Snowmen in partners with one left over (odd)	Snowmen in partners with one left over (odd)	Snowmen in partners because the two leftovers become partners (even)

While students would necessarily need to use a specific number of drawn objects, cubes, or lima beans in their representations, the justifications can be constructed independent of these specific numbers. For example, the sets of interlocking cubes in figure 3.1 illustrate why an odd number plus an odd number equals an even number.

While the initial sets each (necessarily) contain a specific number of cubes—7 and 11—these specific numbers are not critical to the illustration. The important point that is illustrated is that when a group of pairs plus one leftover is added to another group of pairs plus one leftover, the two leftovers combine to form a pair and the end result is a group made up entirely of pairs.

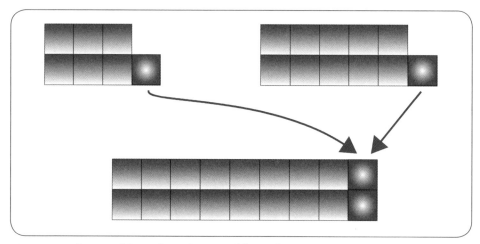

Fig. 3.1. Why an odd number plus an odd number equals an even number

Using Compensation as a Problem-Solving Tool

As we have seen, elementary-school students who are algebraic thinkers have an advantage that extends beyond being well prepared for a more formal study of algebra in later grades. Students who can generalize relationships and reason with those generalizations are also more efficient arithmetic problem solvers.

Let's revisit a previously discussed task (Russell et al. 2011a):

You have 10 vegetables on your plate. Some are peas and some are carrots. How many peas and how many carrots could you have?

Recall that this task can encourage students to write equivalent expressions (e.g., 5 + 5, 9 + 1), express these equivalences as equations (e.g., 5 + 5 = 9 + 1), and strengthen their understanding of the equal sign (SMP 6c). In the context of generalized arithmetic, this task can be extended in another direction to encourage students to look for underlying mathematical structure (SMP 7a).

Suppose a teacher asked students to record the number of peas and the number of carrots they could have. Students might generate something like table 3.1:

Table 3.1. Random solutions for the Peas and Carrots problem

Number of Peas	Number of Carrots
5	5
1	9
7	3
4	6
0	10
9	1

This table is a good first step but does not tackle the problem in a systematic way. A teacher might encourage her students to think more systematically by asking, "How can we be sure we have *all* of the possibilities?" By organizing students' responses in a systematic way, table 3.2 can be constructed as follows:

Table 3.2. All possible solutions for the Peas and Carrots problem organized systematically

Number of Peas	Number of Carrots
0	10
1	9
2	8
3	7
4	6
5	5
6	4
7	3
8	2
9	1
10	0

Using table 3.2, students can make all sorts of observations about the structure they see, for example:

- The Peas column goes from 0 to 10 while the Carrots column goes from 10 to 0.
- The numbers in the Peas column get bigger while the numbers in the Carrots column get smaller.
- When the number in the Peas column goes up by 1, the number in the Carrots column goes down by 1.

Students can also use manipulatives to illustrate that all of the possibilities have been systematically listed. Consider, for example, the following conversation that took place in a first-grade classroom (Russell et al. 2011a, pp. 62–63):

Ms. Callendar: So how did you use . . . 5 plus 5 to help you with the next one?

Cory: Because if you start out with 5 plus 5, then you . . . and then take 1 away from this 5 and add it to this 6, and then this is 4 and so on and so on.

Ms. Callendar:	Can you show me that with these cubes? I am going to give you 5 red cubes and 5 blue cubes for 5 plus 5. Can you show me what you mean, Cory?
Cory:	This is what I mean. I mean, so you pretend you have 5 cubes and 5 cubes [*holding up a stack of 5 red cubes and a stack of 5 blue cubes*]. Then put this 5 onto this [*moving 1 blue cube to the red stack*], and that makes this 6 and this 4 and so on and so on until you get to this [*motioning putting the two stacks together to make 10, implying 10 plus 0*].

Stepping away from the "peas and carrots" context, students might express the generalization "Given two addends, if 1 is subtracted from one addend and 1 is added to the other, the sum remains the same" (Russell et al. 2011a, p. 62), though not with such formal language. Students might instead say "If you are adding two numbers and take 1 away from one of the numbers and give it to the other number, the total stays the same."

The act of identifying and expressing the generalization engages students in important algebraic thinking that will lay the foundation for more formal algebraic thinking in the future. More immediately, however, reasoning with the established generalization assists students as they learn their math facts. If a student knows that $5 + 5 = 10$ but does not yet know $6 + 4 = 10$ to the point of recall, he or she can use algebraic thinking to generate this fact without needing to count.

This concept of compensation comes naturally to children before their work with specific numbers (Britt and Irwin 2011). In a study of children's understanding of operations before being introduced to arithmetic, children were asked what would happen to a total collection of candy divided into two boxes if a doll removed a piece from one of the boxes, added a piece to one of the boxes, moved a piece from one box to the other, or removed a piece from one box while the interviewer added a different piece to the other box. Four-year-old children were certain that the total quantity of candy would stay the same if a piece of candy was moved or replaced but would increase or decrease if the amount in just one of the boxes was altered. Five- and six-year-old children could additionally explain these relationships and express general principles (SMP 2a; PS 3a) in their own language. For example, one student stated that the total after a piece was moved would be "The same, except [the doll] put one of the lollies from here to here" (p. 140). The researchers suggest that it is important to build on what students already understand so that the complexity of learning to understand numbers does not distract them from using their existing knowledge. A task such as the "peas and carrots" problem encourages students to work with numbers while staying grounded in a realistic context.

While this section highlighted specific tasks that provide students with opportunities to generalize mathematical relationships, represent and justify these generalizations, and reason with these generalizations, it is important to remember that opportunities to engage students in these algebraic thinking practices can surface at any time during their day-to-day arithmetic work.

> Whenever a teacher becomes aware of an implicit generality in some particular instances, there is an opportunity to make a choice: to pause for a moment and prompt learners to try to express that generality before continuing, or to keep going. (Mason 2008, p. 80)

These opportunities should be considered while planning lessons and while engaging in work with students. Pausing to shed light on the structure underlying students' arithmetic work is often beneficial.

Functional Thinking

So far in this chapter, we have thought about the importance of equality, expressions, equations, and inequalities and ways of building algebraic thinking in the context of generalizing arithmetic. Another way to foster algebraic thinking in the elementary grades is by having students work with functions. A function is a mathematical statement that describes how two quantities vary in relation to each other. Functional thinking involves building, describing, and reasoning with and about functions (Blanton 2008) and can provide an opportunity for students to engage in aspects of reasoning and sense making that are critical to algebraic thinking: generalizing relationships (SMP 2a); expressing those relationships in words, symbols, tables, or graphs (SMP 2b, c; PS 3a; PS 5a, b); and reasoning with these various representations (SMP 2b, c; PS 2a; Blanton et al. 2011).

Functional thinking often begins by having students examine patterns. It is a rather common activity in the elementary grades to have students find and construct repeating or growing patterns. For example, students might be asked to identify and extend patterns in lists such as those in figure 3.2.

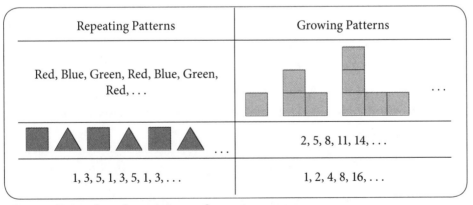

Fig. 3.2 Examples of repeating and growing patterns

While identifying and extending repeating patterns can help students learn to anticipate and look for regularity, this activity does not particularly help develop students' functional thinking. Identifying and extending growing patterns, on the other hand, can engage students in one aspect of functional thinking: looking for *recursive patterns*, or relationships within a sequence. The recursive pattern in the first growing pattern, for example, is "add a square to the top and to the right side." The recursive pattern in the second growing pattern is "plus 3," and the recursive pattern in the third growing pattern is "times 2."

These growing-pattern activities engage students in important thinking but are not by themselves sufficient for developing students' functional thinking. What is missing from these examples is the need to engage in *covariational thinking*, a critical aspect of functional thinking. Covariational thinking involves analyzing how *two* quantities vary *in relation to each other*. In a function, each value of one quantity corresponds in a specific way to a value of a second quantity.

Generalizing from a Specific Case to Build a Function

Functional thinking can be developed by the extension of simple arithmetic tasks. For example, consider the following task (adapted from Blanton and Kaput, 2003):

> At a carnival, you need 10 tickets to win a prize. Jason has already earned 8 tickets. How many more tickets does he need to earn to win a prize?

This problem has a specific answer (two tickets) but can be extended to an algebraic task by varying the total number of tickets needed to win a prize. For example, suppose eleven, twelve, or fifteen tickets are needed. Students might be asked to write a number sentence that describes how many more tickets need to be earned to win a prize in each case. Students might also be asked to generalize the mathematical relationship to describe how the number of tickets that still need to be earned corresponds to the total number of tickets needed to win a prize. This correspondence could be represented in a variety of ways:

In words: The number of tickets that still need to be earned is 8 less than the total number needed to win a prize.

In symbols: $N = T - 8$ where N is the number of tickets that still need to be earned and T is the total number of tickets needed to win a prize.

In a table: See table 3.3.

Table 3.3. Relationship between total number of tickets needed to win a prize and number of tickets that still need to be earned

Total Number of Tickets Needed to Win a Prize	Number of Tickets That Still Need to Be Earned
8	0
9	1
10	2
11	3
12	4
13	5
14	6
15	7
16	8
17	9
18	10
19	11
20	12

Consider another specific task in which students are asked to figure out the total number of eyes that seven dogs have. They might skip count by two or add two repeatedly to find that there are fourteen eyes. How can students generalize the mathematical relationship between the number of dogs and the total number of eyes?

Students in prekindergarten–grade 5 were encouraged to do just that by first being asked to consider how many total eyes an increasing number of dogs would have (Blanton and Kaput 2004). Finding out how many eyes seven or ten dogs have may not be too difficult for some students to accomplish by counting. Asking students to consider a much larger number of dogs encourages them to go beyond counting to find a more efficient way.

What could students do with this task? Blanton and Kaput (2004) found that prekindergarteners spent time with paper cutouts of dogs, counting their eyes and, together with the teacher, using a table to organize their data. For example, they recorded that one dog had two eyes and that two dogs had four eyes. The teacher recorded both the numbers and the corresponding number of dots. Students did not look for patterns or make predictions. They came to their conclusions by counting visible objects. However, the development of a correspondence between numeral and object and the use of a table to organize covarying quantities (PS 5a) were important early steps in developing algebraic reasoning in general and functional thinking in particular.

Kindergarteners likewise recorded their data pictorially, using dots to represent eyes. Teachers again recorded students' data in a table, this time calculating results for up to ten dogs. Some kindergarteners were additionally able to identify the pattern in the number of eyes as "counting by twos" or "more and more." Those who identified the recursive "counting by twos" pattern benefited from engaging in skip counting. Skip counting is important for algebraic thinking because it focuses on relationships among numbers. It is additionally useful for helping students to develop computational fluency and to practice arithmetic skills in an interesting context (Blanton 2008). A few kindergarten students noticed that "every time we add one more dog, we get two eyes" (p. 137). This is a crucial step as it indicates an ability to go beyond skip counting to coordinate two quantities, that is, to engage in covariational thinking (SMP 2a).

By first grade, students were able to record their own data in a table and could describe the pattern as "double" for the total number of eyes. Students in second grade were furthermore able to describe the relationship as "You have to double the number of dogs to get the number of eyes" (p. 138) and could use this relationship to predict the number of eyes for 100 dogs without counting the eyes. Third- through fifth-grade students were additionally able to represent these relationships symbolically and graphically (SMP 2b, c, d; PS 3a; PS 5a, b).

This example shows that a common task with multiple "points of entry" can be used with students across multiple grade and ability levels. Prekindergarten–grade 5 students varied in how deeply they could engage with the task, but all students were able to work with two-variable data and gain experience with important aspects of algebraic reasoning and sense making in the domain of functional thinking.

Generalizing the Relationship Between Two Unknown Quantities to Build a Function

Consider another task in which students are asked to think about the relationship between two unknown quantities (Blanton et al. 2012):

> Carter and Jackson went to a birthday party. They each got a treat bag full of erasers. The bags have exactly the same number of erasers in them. Carter won a game at the party and got 2 more erasers. How would you represent the number of erasers Jackson has? How would you represent the number of erasers Carter has?

Although a specific number of erasers is not given, it can be helpful for students to consider specific amounts in their search for a general rule. They might think, for example, "If Jackson has 4 erasers, how many does Carter have?"

This task was posed to students in kindergarten–grade 2 six weeks into a teaching experiment focused on functional thinking (described in Blanton et al. 2015). In an interview setting, Maria asked Rose, a kindergartener, what she knew about the number of erasers that Jackson and Carter had (Blanton et al. 2012).

Rose: When Carter didn't have the 2 erasers . . . they had the same amount, and now Carter got 2 more than Jackson.

Maria then helped Rose think about specific numbers of erasers before asking her to consider the relationship between the number of Jackson's and Carter's erasers in general.

Maria: Let's suppose Jackson had 1 eraser. How many would Carter have?

Rose: 3.

Maria: 3. OK, I agree. How did you get that?

Rose: Because 1 plus 2 is 3.

Maria: And why did you add 1 plus 2?

Rose: I don't know.

Maria: Where did the 1 come from?

Rose: He had 1 in his bag.

Maria: He had 1 in his bag. And where did the plus 2 come from?

Rose: He won 2 more.

Maria: Let's suppose now Jackson had 2 in his bag. How many would Carter now have?

Rose: I don't know.

Maria: OK, so if there were 2 erasers in each of their bags . . .

Rose: Oh, 4.

Maria: 4. OK. How did you get 4?

Rose: Because when you said there were 2 in each bag; 2 plus 2 is 4.

[Rose did the same for 3 erasers, finding that Carter would have 5.]

Maria then asked Rose how she could organize all of this information. Rose's class had been working on organizing data in a function table (which they called a "t-chart") and, with minimal guidance, she was able to produce the table shown in figure 3.3.

Rose used J to stand for the number of Jackson's erasers and K to stand for the number of Carter's erasers. After Rose recorded several specific cases and the relationship between the number of Jackson's and Carter's erasers, Maria asked Rose to consider unknown quantities:

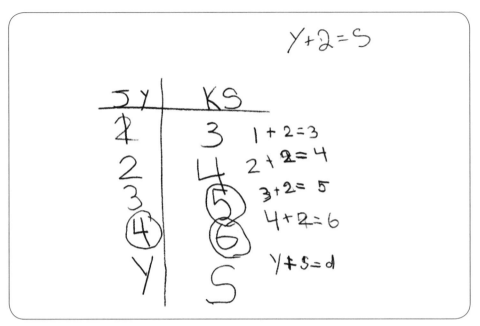

Fig. 3.3. Rose's table

Maria: Do we actually know how many erasers are in this bag? [*Rose shakes her head "no."*] So if we don't know how many we have, what can we do?

Rose's class had been introduced to the idea of using letters to represent unknown varying quantities, so she said she could "use a variable." She suggested that she could call the number of Jackson's erasers *Y* and the number of Carter's erasers *S*.

When asked if she could write an equation to show the relationship between *Y* and *S*, Rose initially wrote $Y + S = d$. With assistance from Maria, Rose was able to see that she should not add the number of Jackson's and Carter's erasers because what was happening every time was the addition of two erasers. Maria wrote the equation $Y + 2 = S$, described as an equation "another student wrote," and asked Rose which equation she thought was correct. Rose immediately said that the new equation was correct because "Carter got two." Maria then asked Rose to think about what would happen if the relationship changed.

Maria: Let's suppose instead of getting two erasers, Carter got 4 erasers. Can you say anything about how your problem changes if you know she's won 4 erasers?

Rose: I would put 4 all the way on that one [*indicating that the series of "+ 2" in the equations to the right of her table would change to "+ 4"*].

Maria: How would your rule change?

Rose: It would be $Y + 4 = S$.

Maria: And tell me why you think it would be that.

Rose: Because instead of 2 she got 4 erasers.

While Rose showed more comfort working with specific values than with a generalized relationship, we see in her work the beginning of several important aspects of functional thinking. She was able to organize data in a function table (PS 5a), engage in covariational thinking (SMP 2a), identify function rules in words (PS 3c), represent relationships using equations (SMP 2b, c, d; SMP 4a, c; PS 5a), and use letters as variables to represent unknown varying quantities (SMP 2b, c). Observations of her work and that of her classmates confirm that even kindergarteners are capable of engaging in rather sophisticated work around functional thinking.

Playing "Guess My Rule" to Build a Function

Another activity that can help students focus on covariational thinking is a game called "Guess My Rule." In this game, a student or the teacher generates several input/output pairs and asks the other students to determine the (functional) rule linking the two sets of numbers. For example, table 3.4 illustrates the "add 4" rule:

Table 3.4. The "add 4" rule

Input	Output
4	8
3	7
8	12

The nonsequential nature of the table encourages students to focus on the "across" or functional rule rather than on the "down" pattern or "what comes next" way of thinking (Moss and McNab 2011). Furthermore, students need to test several data points to determine the rule. For example, if only considering the first data point, a student might conclude that the rule is "times 2."

Generalizing Geometric Growing Patterns to Build a Function

There is growing evidence that having students engage with geometric growing patterns along with numeric ones can develop students' functional thinking (Moss and McNab 2011). Unlike the geometric pattern presented in the introduction to this section, in which the focus is on finding the "next" picture in a series, this work focuses on finding the functional relationship or correspondence between the "position number" and the number of elements in each position. Consider, for example, the tile pattern in figure 3.4, where the number beneath each group of tiles indicates the position number:

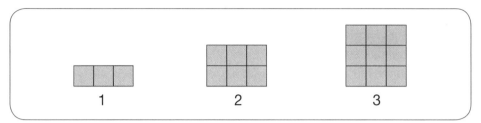

Fig. 3.4 Growing tile pattern

Students can be asked questions about this pattern such as the following:

- If this pattern keeps growing in the same way, what would the next position look like?
- How many blocks would be in the next position?
- What would the tenth position look like?
- How many blocks would be in the tenth position?
- How many blocks would be in the hundredth position?

While students might initially tend to continue the pattern to find the number of blocks in the next and perhaps even the tenth position, asking students to find the number of blocks in the hundredth position encourages them to find an explicit relationship linking the position number to the number of blocks. In this case, the position number times 3 gives the number of blocks, so there would be 300 blocks in the hundredth position.

The example can be extended to encourage students to find two-part function rules (see fig. 3.5).

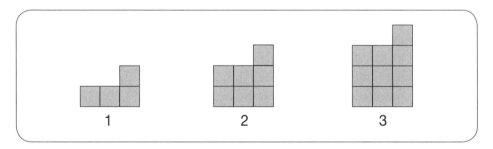

Fig. 3.5 Growing tile pattern described by a two-part function rule

In this case, the position number times 3 plus 1 gives the number of blocks. Second graders who worked with these patterns referred to the constant, 1, as the "bump" (Moss and McNab 2011). In more formal algebra, students might refer to 3 as the slope (i.e., the rate of change) and 1 as the *y*-intercept (i.e., the number that would correspond to position 0).

Functions are important tools that help us make sense of data and the world around us. The study of functions is, unfortunately, often reserved for the late middle school or high school years. However, given the appropriate tasks, children as young as prekindergarten can work with two-variable data, and children as young as kindergarten can begin to engage in covariational thinking (SMP 2a). Functions can furthermore serve as a context for students to develop key elements of algebraic reasoning and sense making. Students are generalizing mathematical relationships when they discuss how two sets of data are related (SMP 2a); they are representing generalizations when they express function rules in words, tables, or symbolic notation (SMP 2b, c, d; PS 3a, c; PS 5a); and they are reasoning with generalizations when they use function rules to predict data points far into the future (e.g., the total number of eyes one hundred dogs would have). Tasks used in everyday instruction can often be extended to provide students with opportunities to develop functional thinking.

Variables

When asked to consider the domain of algebra, many of us immediately think of variables. High school algebra courses are often remembered as places spent manipulating variable expressions with little concern for what the variables represented. It is indeed the case that part of algebra's power comes from the ability to work with symbolic representations without constantly paying attention to what the symbols represent. A problem surfaces, however, when what the symbols stand for is so de-emphasized from the very beginning of students' experiences that meaningful connections between the symbols and their real-life referents are never made. It is critical that this connection be maintained for prekindergarten–grade 2 students working with expressions and equations and that students continue to use variables as symbolic tools to represent ideas they already understand.

Thus far in this chapter, we have seen examples of students engaged in algebraic thinking around the big ideas of equality, expressions, equations and inequalities; generalized arithmetic; and functional thinking. Some of their work in representing and reasoning with generalizations has involved the use of variables. In this section we will take a closer look at the big idea of variable, its different uses in the elementary grades, and some of the key understandings associated with this big idea that young students can develop.[2]

Variables as Specific Unknowns

Traditional high school algebra courses emphasize the use of variables to represent fixed but unknown values. For example, in the equation $y + 9 = 12$, there is only one value of y that will make the equation true. By solving the equation, we can determine that value, $y = 3$.

Where might prekindergarten–grade 2 students gain experience with this use of variables? In the Equality, Expressions, Equations, and Inequalities section of this chapter, we discussed the use of open number sentences to develop students' understanding of the equal sign. In the equation $8 + 4 = \square + 5$, the box stands for a specific unknown value. It is not the case that any number can be placed in the box to make the number sentence true. Only when a 7 is placed in the box is the number sentence true. Researchers have found that students generally begin their work with unknowns with open number sentences, using boxes or blanks to "fill in." The subtle shift to thinking "What number can we *replace* the box with to make the number sentence true?" enables students to readily adapt to using letters to represent unknowns (Carpenter et al. 2003).

One opportunity for students to gain experience writing algebraic equations and solving for variables is to extend their arithmetic problem-solving activities to include this work. Consider the following previously discussed task:

> *If Evelyn had 2 books and borrowed 6 more books from the library, how many books does she have now?*

As discussed, this task provides the opportunity to engage students in thinking about the commutative property of addition. The task can be further extended by asking students to represent the problem situation with an equation, using either a box or a letter to represent the unknown value (e.g., $2 + 6 = a$).

Solving for the value of a variable in the early grades should not resemble the rule-focused approaches many of us experienced in formal algebra courses. Students should instead be encouraged to take the time to inspect equations, examine their structure, and think about their meanings. For example, given the equation $c + 6 = 14$, a student might think, "That means some number plus 6 is the same as 14. I need to figure out what that number is." The student might count on from 6 to 14 or subtract 6 from 14. What is important is that the strategy chosen make sense to the student (SMP 1c, g) and not be an imposed rule (e.g., subtract 6 from both sides) that is applied without understanding.

Younger students who do not yet have experience writing number sentences might play the game "How many are under the cup?" (Greenes et al. 2001). In this game, one student counts out five chips and places them under a cup. Another student takes some chips from under the cup and counts them. A third student answers the question "How many chips are left under the cup?" Students might first represent their findings with drawings of a cup and chips inside and outside the cup. Over time they might represent their findings in a three-column table: number of chips in the cup, number of chips taken out, number of chips left under the cup (PS 5a). Eventually, they might write number sentences such as $2 + \square = 5$ or $5 - 2 = \square$ to represent their data (SMP 4a, c). This is yet another example of a task that both engages students in algebraic thinking and strengthens their arithmetic fluency.

Variables as Varying Quantities

Another use of variables that students in the elementary grades should encounter is that of variables as varying quantities. One way to introduce this use of variables is to take a typical arithmetic task and transform it by varying one of the parameters in the task (Blanton, 2008). We saw this in the discussion of functional thinking when students were asked to generalize the mathematical relationships present in the Carnival, Dogs and Eyes, and Erasers tasks. For example, in Dogs and Eyes, the arithmetic task of finding the total number of eyes on a specific number of dogs is transformed when one varies the number of dogs and looks for a relationship between the number of dogs and the total number of eyes.

Consider another example. After solving the problem "If Samantha has five stickers and Brady has three more stickers than Samantha, how many stickers does Brady have?" students might be asked the following, where Samantha's known number of stickers (five) is now unknown:

> If Samantha has some stickers and Brady has 3 more stickers than Samantha, how would you represent the number of stickers Brady has?

This new problem does not have a numerical answer. The number of stickers Brady has must be represented in a different way. This can be awkward for children who have only worked with arithmetic tasks with numerical answers, but after some experience they can become comfortable with the ambiguity present in the task. Students can be helped to think about how to represent the number of stickers Brady has by considering questions such as the following:

- What do you know about the number of stickers Samantha has?
- How should we describe this number of stickers?
- Does Brady have more or fewer stickers than Samantha?
- How should we describe the number of stickers Brady has?

Students might express the number of stickers Brady has as "the number of stickers Samantha has plus three" or as "three more than the number of stickers Samantha has." With guidance, they can use a variable to represent the varying number of stickers Samantha could have and represent the number of stickers Brady has as $s + 3$, where s represents the number of stickers Samantha has (SMP 4a, c; PS 5a).

Another effective activity in the early grades to encourage the use of variables as varying quantities is to find different ways to partition a number into pairs of addends (Greenes et al. 2001). Let's revisit the Carrots and Peas task:

You have 10 vegetables on your plate. Some are peas and some are carrots. How many peas and how many carrots could you have?

As discussed, this task encourages students to systematically consider a wide range of possible solutions, organize data, and generalize mathematical relationships and structure. It can also be extended to provide a context for students to work with variables as varying quantities. When pushed to think of an equation that could encompass all of the possibilities, students might suggest that the situation could be represented by $\square + \Delta = 10$. When children are taught to represent unknown, varying quantities with letters, they might suggest $p + c = 10$.

Unlike the case of equations such as $c + 6 = 14$ and $2 + \square = 5$, here there is more than one value the variables may take on. Students who pose just one possibility (e.g., $p = 2$ and $c = 8$) should be encouraged to think of others.

There are a few common misconceptions that many older students have about variables that this task can encourage students to confront. First, many students treat variables as labels or abbreviations rather than as letters that stand for quantities (e.g., Booth 1988). This is understandable given the way letters are often used. We may use $2m$ to stand for 2 meters and $3f = 1y$ to indicate that 3 feet is the same as 1 yard. Building on this experience, students might think of c as standing for "carrots" rather than "the number of carrots." This distinction is important. If, for example, we let f stand for the number of feet and y stand for the number of yards, the equation $3f = 1y$ is no longer correct. This misconception is more likely to surface when the letters chosen to represent the quantities match the first letters of the objects (McNeil et al. 2010) as in the case of choosing c for the number of carrots and p for the number of peas.

Another more easily addressed misconception that some students have is that different letters cannot stand for the same number. For example, when asked whether the equation $c = r$ was true or false, many middle school students thought it must be false because, as one sixth-grade student stated, "When a letter represents a number, usually each letter represents a different number, not the same ones" (Stephens 2005, p. 97). The fact that c and r can in fact be equivalent is a mathematical convention, not a notion that is intuitively obvious. The Carrots and Peas task is a problem situation that supports the adoption of this convention because it makes sense that someone could have 5 carrots and 5 peas and that therefore in the equation $c + p = 10$, c and p could have the same value.

Variables as Generalized Numbers

Finally, elementary students should be introduced to the use of variables as generalized numbers. In the earlier discussion of generalized arithmetic, we saw examples of students generalizing mathematical relationships concerning the fundamental properties and relationships between operations. Variables—expressed as boxes or letters—can be used to express these relationships in succinct ways. Consider again the students who came to express the zero property of addition—"a number plus zero equals that number"—as $m + 0 = m$. They could likewise represent the commutative property of addition as $a + b = b + a$.

In these cases, variables are used not to represent specific unknowns or quantities that vary in a problem-solving context but rather mathematical situations that are always true. While students generally begin by representing these relationships with words before representing them with variables, after they become comfortable with variable notation, they may find that some relationships are more easily expressed symbolically (Carpenter et al. 2003). For example, $a + b - b = a$ is a combination of two fundamental properties ($b - b = 0$ and $a + 0 = 0$) and may be easier for students to represent with symbols than with words.

In this chapter, we have seen examples of students working with specific numbers in general ways. Alicia used specific numbers of blocks to justify the commutative property of addition and students in a first-grade classroom used specific numbers of lima beans to generalize relationships concerning even and odd numbers, for example, with the full understanding that their choices of numbers were arbitrary. There are other ways to help students bridge the gap between working with specific numbers and working with variables. Consider the following true/false and open number sentences:

True/False Number Sentences	Open Number Sentences
$23 - 14 + 14 = 23$	$17 + 5 - 5 = \square$
$12 + 15 = 15 + 12$	$8 + 14 = \square + 15$

Students who provide general explanations of why number sentences like $23 + 14 - 14 = 23$ are true are said to be engaging in *quasi-variable* thinking (Fujii and Stephens 2008). They may not yet be able to grasp that the variable a in $23 - a + a = 23$ could represent a wide variety of numbers, including rational and negative numbers. Nonetheless, they are essentially able to treat the 14 as a variable and recognize that it does not matter what number is in that particular position. The 14 could be any number, and the equation would remain true. After seeing several instantiations of the $23 - a + a = 23$ relationship with numbers, students might eventually make sense of the variable representation.

Treating numbers as quasi variables can help students solve all sorts of arithmetic computations, even those less obvious. A student may compute the sum $73 + 19$, for example, by first subtracting 1 from 73 and adding it to 19 to produce the equivalent but easier sum $72 + 20$ (SMP 1c, d, h; SMP 7a, c; PS 1a)

(Blanton et al. 2011). This student is employing the algebraic relation $a + 19 = (a - 1) + 20$ or, more generally, $a + b = (a - c) + (b + c)$, even though the formal notation is not used.

In traditional, arithmetic-based elementary classrooms, students are typically exposed to variables only as unknowns (e.g., in "fill in the-box" number sentences). As several examples in this section have shown, however, it is not unreasonable to expect young students to work with variables as varying quantities as well as generalized numbers. Notice, furthermore, that in all of these examples, variables are used to represent ideas that already make sense to students. They are not asked to apply rules for manipulation that do not make sense to them. As Smith and Thompson (2008) argue, "If . . . students develop mathematical ideas of sufficient complexity—among them complex quantities and relationships between quantities—their expression, manipulation, and further abstraction in algebraic notation can become a more meaningful and sensible activity" (p. 127).

Conclusion

This chapter addressed four "big ideas" from the domain of algebra in which students in prekindergarten–grade 2 can engage: equality, expressions, equations, and inequalities; generalized arithmetic; functional thinking; and variables. Within each of these big ideas, we shared examples of tasks that have the potential to elicit key elements of algebraic reasoning—generalizing mathematical relationships and structure, representing generalizations, justifying generalizations, and reasoning with generalizations—as well as evidence that young students can engage in these practices. Opportunities exist in everyday instruction to engage students in these important algebraic thinking practices as they relate to these big ideas. Doing so can strengthen students' understanding of arithmetic, lay an important foundation for their future, more formal study of algebra, and instill in students the conviction that mathematics is and should be a reasoning and sense-making activity.

Endnotes

1. See Blanton, Levi, Crites, and Dougherty (2011) for a complete list and discussion of the fundamental properties.
2. See Blanton, Levi, Crites, and Dougherty (2011) for further discussion of the roles of variables.

References

Baroody, A. J., and H. P. Ginsburg. "The Effects of Instruction on Children's Understanding of the 'Equals Sign.'" *Elementary School Journal* 84, no. 2 (1983): 199–212.

Bastable, V., and D. Schifter. "Classroom Stories: Examples of Elementary Students Engaged in Early Algebra." In *Algebra in the Early Grades*, edited by J. J. Kaput, D. W. Carraher, and M. L. Blanton, pp. 165–184. New York: Lawrence Erlbaum, 2008.

Blanton, M. L. *Algebra and the Elementary Classroom: Transforming Thinking, Transforming Practice.* Portsmouth, NH: Heinemann, 2008.

Blanton, M., B. Brizuela, A. Gardiner, K. Sawrey, B. Gravel, and A. Newman-Owens. "Analyzing Learning Trajectories in Grades K–2 Children's Understanding of Function." Unpublished raw data, 2012.

Blanton, M., B. Brizuela, A. Gardiner, K. Sawrey, B. Gravel, and A. Newman-Owens. "A Learning Trajectory in Six-year-olds' Thinking About Generalizing Functional Relationships." *Journal for Research in Mathematics Education* 46, no. 5 (2015): 511–558.

Blanton, M. L., and J. J. Kaput. "Developing Elementary Teachers' 'Algebra Eyes and Ears.'" *Teaching Children Mathematics*, 10, no. 2 (2003): 70–77.

Blanton, M. L., and J. J. Kaput. (2004). "Elementary Grades Students' Capacity for Functional Thinking." In *Proceedings of the 28th PME International Conference, vol. 2*, edited by M. J. Hoines and A. B. Fuglestad, pp. 135–142.

Blanton, M. L., L. Levi, T. Crites, and B. J. Dougherty. *Developing Essential Understandings of Algebraic Thinking, Grades 3–5.* Reston, VA: The National Council of Teachers of Mathematics, 2011.

Booth, L. R. "Children's Difficulties in Beginning Algebra." In *The Ideas of Algebra, K–12*, edited by A. Coxford and A. Schulte, pp. 20–32. Reston, VA: National Council of Teachers of Mathematics, 1988.

Britt, M. S., and K. C. Irwin. "Algebraic Thinking With and Without Algebraic Representation: A Pathway for Learning." In *Early Algebraization: Curricular, Cognitive, and Instructional Perspectives* edited by J. Cai and E. Knuth, pp. 137–159. Heidelberg, Germany: Springer, 2011.

Carpenter, T. P., E. Fennema, M. L. Franke, L. Levi, and S. B. Empson. *Children's Mathematics: Cognitively Guided Instruction.* Portsmouth, NH: Heinemann, 1999.

Carpenter, T. P., M. L. Franke, and L. Levi. *Thinking Mathematically: Integrating Arithmetic and Algebra in the Elementary School.* Portsmouth, NH: Heinemann, 2003.

Carpenter, T. P., and L. Levi. "Developing Conceptions of Algebraic Reasoning in the Primary Grades." National Center for Improving Student Learning and Achievement in Mathematics and Science, University of Wisconsin, Madison, Wis., 2000. http://ncisla.wceruw.org/publications/reports/RR-002.PDF

Dougherty, B. "Measure Up: A Quantitative View of Early Algebra." In *Algebra in the Early Grades*, edited by J. J. Kaput, D. W. Carraher, and M. Blanton, pp. 389–412. New York: Lawrence Erlbaum, 2008.

Fonger, N., A. Stephens, M. Blanton, and E. Knuth. "A Learning Progressions Approach to Early Algebra Research and Practice." In *Proceedings of the 37th Annual Meeting of the North American Chapter of the International Group for the Psychology of Mathematics Education*, edited by T. G. Bartell, K. N. Bieda, R. T. Putnam, K. Bradfield, and H. Dominguez, pp. 201–204. East Lansing, MI: Michigan State University, 2015.

Fujii, T., and M. Stephens. "Using Number Sentences to Introduce the Idea of Variable." In *Algebra and Algebraic Thinking in School Mathematics*, edited by C. E. Greenes and R. Rubenstein, pp. 127–140. Reston, VA: National Council of Teachers of Mathematics, 2008.

Greenes, C., M. Cavanagh, L. Dacey, C. Findell, and M. Small. *Navigating Through Algebra in Prekindergarten–Grade 2.* Reston, VA: National Council of Teachers of Mathematics, 2001.

Hattikudur, S., and M. W. Alibali. "Learning About the Equal Sign: Does Comparing with Inequality Symbols Help?" *Journal of Experimental Psychology* 107(2010): 15–30.

Knuth, E. J., A. C. Stephens, N. M. McNeil, and M. W. Alibali. "Does Understanding the Equal Sign Matter? Evidence from Solving Equations." *Journal for Research in Mathematics Education* 37, no. 4 (2006): 297–312.

Mason, J. "Making Use of Children's Powers to Produce Algebraic Thinking." In *Algebra in the Early Grades*, edited by J. J. Kaput, D. W. Carraher, and M. L. Blanton, pp. 57–94. New York: Lawrence Erlbaum, 2008.

McNeil, N. M., A. Weinberg, S. Hattikudur, A. C. Stephens, P. Asquith, E. J. Knuth, and M. W. Alibali. "A is for Apple: Mnemonic Symbols Hinder the Interpretation of Algebraic Expressions." *Journal of Educational Psychology* 102, no. 3 (2010): 625–634.

Moss, J., and S. L. McNab. "An Approach to Geometric and Numeric Patterning That Fosters Second Grade Students' Reasoning and Generalizing About Functions and Co-variation." In *Early Algebraization: Curricular, Cognitive, and Instructional Perspectives*, edited by J. Cai and E. Knuth, pp. 277–301. Heidelberg, Germany: Springer, 2011.

National Council of Teachers of Mathematics. *Principles and Standards for School Mathematics.* Reston, VA: The National Council of Teachers of Mathematics, 2000.

National Governors Association for Best Practices (NGA Center) and Council of Chief State School Officers (CCSSO). *Common Core State Standards for Mathematics. Common Core State Standards (College- and Career-Readiness Standards and K–12 Standards in English Language Arts and Math).*Washington, D.C.: NGA Center and CCSSO, 2010. http://www.corestandards.org.

Russell, S. J., D. Schifter, and V. Bastable. *Connecting Arithmetic to Algebra.* Portsmouth, NJ: Heinemann, 2011a.

Russell, S. J., D. Schifter, and V. Bastable. "Developing Algebraic Thinking in the Context of Arithmetic." In *Early Algebraization*, edited by J. Cai and E. Knuth, pp. 43–69. Heidelberg, Germany: Springer, 2011b.

Schifter, D. "Reasoning About Operations: Early Algebraic Thinking in Grades K–6." In *Developing Mathematical Reasoning in Grades K–12*, edited by L. V. Stiff, pp. 62–81. Reston, VA: National Council of Teachers of Mathematics, 1999.

Schifter, D., V. Bastable, S. J. Russell, L. Seyferth, and M. Riddle. "Algebra in the Grades K–5 Classroom: Learning Opportunities for Students and Teachers." In *Algebra and Algebraic Thinking in School Mathematics*, edited by C. E. Greenes and R. Rubenstein, pp. 263–277. Reston, VA: National Council of Teachers of Mathematics, 2008.

Smith, J. P., and P. W. Thompson. "Quantitative Reasoning and the Development of Algebraic Reasoning." In *Algebra in the Early Grades*, edited by J. J. Kaput, D. W. Carraher, and M. L. Blanton, pp. 95–132. New York: Lawrence Erlbaum, 2008.

Stephens, A. C. "Developing Students' Understandings of Variable." *Mathematics Teaching in the Middle School* 11, no. 2 (2005): 96–100.

Understanding and Developing Young Children's Reasoning and Sense Making in Decomposing Geometric Shapes[1]

Michael T. Battista

Katy, a second grader, was shown that a plastic inch-square fit in the upper-left indicated square on the 7-by-3-inch rectangle displayed in figure. 4.1a. She was then asked, "How many plastic squares just like this would it take to completely cover this rectangle?" Katy drew and counted 30 squares, as shown in figure. 4.1b (Battista 1999).

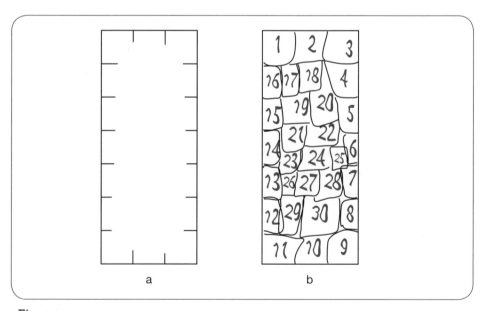

a   b

Fig. 4.1

Katy was then asked to predict how many plastic squares were needed to cover the rectangle shown in figure. 4.2a. Without drawing, Katy pointed and counted as in 4.2b, predicting 30. When checking her answer with plastic squares, she pointed to and counted squares as shown in figure 4.2c, getting 30. When she counted the squares again, first she got 24, then 27.

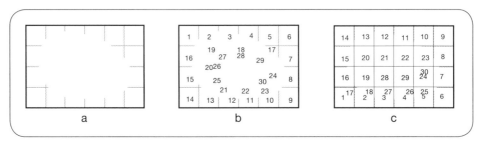

Fig. 4.2

Katy was not using row-by-column organizational reasoning to locate squares in these rectangular arrays. Although educated adults instantly "see" the squares arranged by rows and columns, Katy had not yet mentally constructed this organizing structure in her drawings or counting, even when she used physical materials.

Katy's difficulty is not unusual. When the problem in figure 4.2a was given to above-average students in grades 2–5, only 19 percent of second graders, 31 percent of third graders, 54 percent of fourth graders, and 78 percent of fifth graders made correct predictions (Battista 1999).[2] Clearly, early mathematics instruction is not enabling students to become proficient reasoners about decomposing shapes into organized sets of unit-squares in ways that are essential for measuring area.

In this chapter, we examine the reasoning and sense making that occur as young children decompose and compose interiors of geometric shapes, and how instruction can support the development of this reasoning. We focus on geometric decomposition because almost all of the Common Core geometry and geometric measurement standards for kindergarten–grade 8 involve decomposition in some way.[3] Moreover, geometric decomposition is also important because it involves spatial reasoning. As *Principles and Standards for School Mathematics* (*Principles and Standards*) states, "Spatial visualization—building and manipulating mental representations of two- and three-dimensional objects and perceiving an object from different perspectives—is an important aspect of geometric thinking" (NCTM 2000, p. 41). Almost all geometric reasoning, sense making, and problem solving are intimately connected to spatial reasoning. Furthermore, the National Research Council states that "[u]nderpinning success in mathematics and science is the capacity to think spatially" (2006, p. 6), a statement backed by much research (Newcombe 2010; Wai et al. 2009). And, because focusing on spatial reasoning in mathematics can improve the attitudes and self-confidence of low-achieving middle school students (Wheatley and Wheatley 1979), it can also make mathematics accessible to more students. Indeed, one of the primary ways that we make sense of things is to visualize them. So problem solving that involves spatial reasoning provides a powerful instructional context for developing ALL students' overall mathematical reasoning and sense making.

Processes and Practices in Geometric Reasoning and Sense Making

There are two fundamentally important mental processes that support reasoning and sense making in decomposing and composing geometric shapes: visualizing and spatial structuring. Both processes, which are explained below, are crucial for implementing the *Common Core State Standards for Mathematics* Standards for Mathematical Practices (SMP) and the National Council of Teachers of Mathematics' Process Standards (PS).[4] Visualizing and structuring enable students to make sense of problems (SMP 1); break problems into parts (SMP 3c); find relationships (SMP 1a); make sense of quantities (SMP 2a); use number to model geometric quantities (SMP 4a); look for and make use of structure (SMP 7); notice regularity (SMP 8); solve problems (PS 1); reason about spatial relationships (PS 2); organize, analyze, and evaluate geometric thinking (PS 3); interconnect geometric concepts and connect geometric concepts to number concepts (PS 4); and mentally represent geometric and spatial thinking and objects (PS 5).

Spatial Visualization

Visualization is the process of creating and manipulating mental images so that we can analyze, make sense of, and reason about spatial objects and actions even when they are not visible. Visualization enables us to reason about geometric *representations,* such as physical objects, pictures/diagrams, and computer animations (SMP 1f, 1i; SMP 2d; SMP 3d; SMP 6c; PS 5). Students develop visualization skills as they internalize acting and reflecting on physical objects, pictures, and visual animations.

Spatial Structuring

To *spatially structure* an object or set of objects is to mentally organize it by identifying its component parts and determining how these parts are put together to form the whole (SMP 1b, 2a, 7d). Katy and many other students made errors on rectangle-covering problems because they had not yet mentally constructed a row-by-column structuring that properly located and organized unit-squares within the rectangles. Spatial structuring is critical to making sense of geometric quantities and their relationships (SMP 2a), and it is the foundation for *meaningful* geometric measurement (Battista 2007; Battista et al. 1998). For instance, students who spatially structure a rectangular array of squares into rows or columns usually correctly enumerate the squares by skip counting or multiplying, whereas students who do not employ row or column structuring count squares by ones, often inaccurately, as did Katy.

A critical component of spatial structuring is locating parts of shapes with respect to other parts. For instance, one shape might be located *above* or *to the right of* another shape. Or, to locate Square X in the array shown in figure 4.3a, a student might "see" the square informally in terms of a two-dimensional coordinate-like system. For example, it is in the fourth column to the right *and* the second row down.

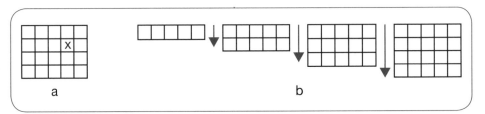

Fig. 4.3

Another component of spatial structuring is *organizing-by-composites*, which combines basic units into more complicated *composite units* that are repeated or iterated to generate a larger shape. (For brevity, the term *composite unit*, which is a unit consisting of more basic units, will be shortened to *composite*.) For instance, to enumerate unit-squares in a rectangle, students might mentally unite squares in a row to form a row-composite that they iterate in the direction of a column to generate the whole array (fig. 4.3b). Similarly, to find the volume of a rectangular box, a student might iterate a composite unit consisting of a layer of unitcubes.

In addition to decomposing shapes for measurement, there are numerous other kinds of spatial structuring that are critical to geometric reasoning. For instance, we distinguish different types of quadrilaterals by decomposing them into sides and specifying their structure by describing relationships between the sides. Parallelograms have "opposite sides equal," whereas kites have pairs of "adjacent sides equal" (fig. 4.4a). Similarly, we analyze and structure three-dimensional shapes by decomposing them into edges, vertices, and faces. For instance, we can characterize a right rectangular prism (a rectangular box) by decomposing it into opposite faces that are congruent rectangles (fig. 4.4b). (We can further structure such a prism by noting that opposite faces are slide/translation images of each other.) This type of analysis eventually leads to understanding properties of shapes (SMP 2d) and to seeing regularity in shapes and combinations of shapes (SMP 8b).

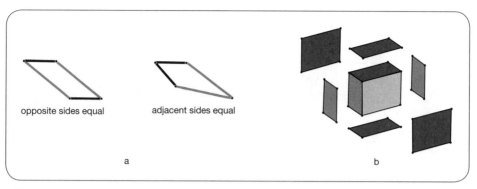

opposite sides equal   adjacent sides equal

a   b

Fig. 4.4

The Key to Visualizing and Structuring: Mental Models

Both visualization and spatial structuring are accomplished by forming and using mental models (Battista 2007). Mental models are nonverbal image-like versions of situations that have the structures of the situations they represent. We understand or make sense of situations when we construct, activate, and interact with appropriate mental models (Johnson-Laird 1983). For instance, we each have a mental model that enables us to navigate from our home to the school where we teach, to imagine things we might see along the route, and to determine detours if our normal route is blocked by construction. When we activate a mental model, we can mentally visualize moving around in it, or moving, combining, and transforming objects in it, just like we do with objects in the physical world. Mental models capture the structure we mentally construct for objects, and when we visualize, we are mentally viewing our mental models. We can help students develop proficiency in using geometric mental models by having them manipulate spatial objects and reflect on those manipulations.

Drawings as Windows on Students' Mental Models

Katy's drawings give us clear glimpses of the mental models she used to reason about enumerating squares in rectangles. Often drawings expose reasoning and structuring that is not as sophisticated as might be assumed from students' correct answers. For example, Branford was shown a 3-by-4-inch rectangle and was asked to cover it with square tiles. He correctly counted the squares as he placed them in the rectangle in the order shown in figure 4.5a. The teacher, who was trying to help Branford better structure arrays of squares, then covered the squares and asked Branford to draw them. Branford's drawing indicated that his mental model of the squares was not yet properly structured into rows and columns (fig. 4.5b). When the teacher uncovered the square tiles and asked Branford if his drawing was correct, Branford said that it was.

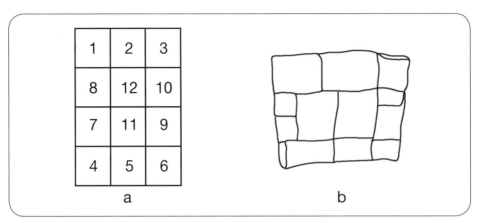

Figs. 4.5

Branford structured the array into top and bottom rows, partial right and left sides, and "2 squares in the middle." However, his unit of "2 squares in the middle" was incorrectly turned 90°. Although Branford's mental model of the array only partially captured its row-by-column structure, his partial structuring is an important step in his development of a complete row-by-column structuring.

A Learning Progression for Visualizing Shape Decompositions

For numerous mathematical topics, researchers have found that students' development of mathematical knowledge and reasoning can be characterized in terms of "levels of sophistication" (Battista 2011, 2012a, b; Clements and Sarama 2009). A *level of sophistication* is a distinct type of conceptualizing and reasoning that occurs within a hierarchy of reasoning levels for a mathematical topic. A set of levels of sophistication forms a *learning progression*, which is a description of the successively more sophisticated ways of reasoning and sense making that students pass through in developing deep understanding of a topic.[5] Levels of sophistication can be thought of as mental plateaus that students ascend in a learning progression (fig. 4.6).

In the remaining sections of this chapter, we examine levels of sophistication in young students' visualization and decomposition reasoning for three types of instructional tasks. These levels are critical to understanding students' reasoning and sense making in a way that enables us to adjust teaching to meet their learning needs. Embedded within the levels discussions, and in Appendixes C and D, are instructional tasks that are specifically designed to help students ascend to higher levels, which should be the goal of instruction. Students should master reasoning about tasks at one level before moving on to tasks at the next level. The number of tasks of each type needed before students master the

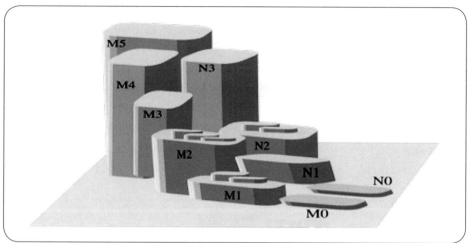

Fig. 4.6

reasoning required by the tasks varies by student. There are also descriptions of the interactional guidance teachers must give to students to support their sense making and reasoning about these tasks.

Because the instructional tasks are based on learning progressions, most students can learn effectively by solving appropriate problems (PS 1b); that is, when given problems that are within their 'cognitive reach' according to a learning progression, students create new ideas and reasoning that meaningfully extend or supplant their current ideas. It is important that this type of instruction constantly encourages students to make sense of ideas and reasoning and to analyze and evaluate their own thinking and that of their peers (PS 3b). In this instructional approach, students build new ideas from the ideas they already have, so they are constantly interconnecting mathematical ideas (PS 4a).

Note that part of describing student reasoning at a particular level is analyzing what students can and cannot do with that particular form of reasoning. This approach is not a "deficit" model of students' reasoning; instead we are trying to understand different types of student reasoning well enough so that we can choose sets of instructional tasks that enable students to build on and revise their current ways of reasoning. With proper instructional guidance, individual students will gradually overcome the limitations of their current ways of reasoning to develop the kinds of formal reasoning required in mathematics curricula.

One way to encourage and support students' development of the visualization needed for shape decomposition and analysis is to have them solve pattern-block frame tasks. These tasks can be implemented by using physical pattern blocks or virtual blocks in a dynamic geometry environment.[6] After examining examples of frame tasks, we investigate levels of sophistication in students' reasoning about decomposing shapes in these tasks. We also discuss instructional activities that promote students' development of this critically important reasoning.

Frame Tasks: Abstracting, Visualizing, and Decomposing Pattern-Block Shapes

In frame tasks, students use pattern blocks (fig. 4.7) to completely cover the inside of shape "frames" or outlines (figs. 4.8a and 4.8b). (The frames should be the same size as the pattern blocks so that students can find or check solutions by physically manipulating the blocks.) Of course, Tangram tasks can also be used.

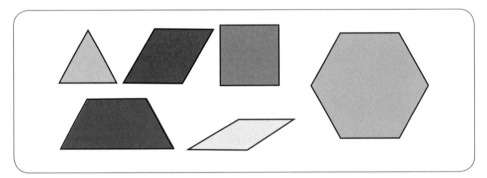

Fig. 4.7

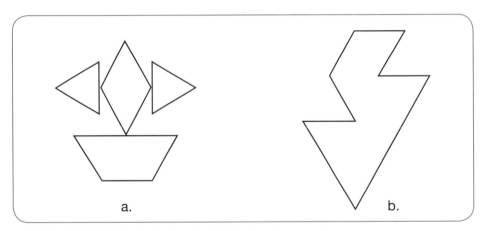

Fig. 4.8

Levels of Sophistication in Students' Reasoning About Decomposing Shapes[7]

Level 0

The student uses holistic, approximate shape matching. Students reason about shapes only as visual wholes. Their shape matching is often approximate or inaccurate, and they cannot decompose larger shapes into smaller shapes even when they have physical shapes to manipulate.

Example 4.1

For the frame task shown in figure 4.8a, Jason placed pattern blocks as shown in figure 4.9. Because his mental representations of the shapes were vague, his decomposition of the frame was approximate, not precise (SMP 6). Indeed, he correctly identified the triangles, but his placement of the two light blue triangles was off; he used a light gray rhombus instead of the correct dark blue rhombus in the top row; and he did not recognize the bottom shape as a trapezoid, or a dark blue rhombus combined with a light blue triangle.

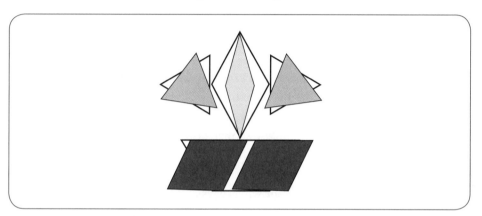

Fig. 4.9. Jason's work

Level 1

The student uses precise holistic shape matching through physical trial and error: there is a beginning recognition of physical decomposition. Students reason more precisely about shapes as visual wholes (SMP 6). Although their visualization of shape placements is approximate, they can correctly determine shape placements by using trial and error to physically manipulate pattern blocks.

Students start to recognize the possibility that some shapes can be made from other shapes (e.g., two light blue triangle pattern blocks can make the dark blue rhombus pattern block). But they recognize such a decomposition only after they have completed it physically. If asked if the trapezoid can be made by other shapes, they use random trial and error in an attempt to determine the answer.

Example 4.2

For the frame task shown in figure 4.8a, Janine placed a light blue triangle on the triangle on the right, then slowly rotated it until it fit. She followed that up by doing the same thing for the triangle on the left. She then tried to place the light gray rhombus in the center, found it didn't work, then tried the dark blue rhombus, again rotating it until it fit (not anticipating the correct orientation). Finally she tried the dark gray trapezoid on the bottom trapezoid, but she placed it upside down. At first, she thought that maybe the dark blue rhombus fit there. But after trying a dark blue rhombus, she returned to the dark gray trapezoid, slowly rotating it until it fit properly.

Instruction to Help Students Move from Level 0 to Level 1

Task sheets like Types 1–4 in Appendix C should be given to students one at a time, in order, directing students to: "Try to make the shape with the blocks." For each problem, students should have ALL the pattern blocks, so that they have to decide which shapes to use. Students who struggle should be given the correct set of pattern blocks to use. For students who still struggle, first show them the solution then remove it to see if they can repeat it. The goal is for students to accurately match shapes while using physical blocks. We should always check to see if students position pattern blocks correctly on the shape pictures.

Level 2

The student accurately visualizes matching shapes and physically uses decomposition. Because students' shape imagery starts to account for the measures of shapes' sides and angles, their imagery becomes more powerful and accurate. Students can mentally visualize how a given shape might fit into a shape frame if the outline of the given shape is fairly clear (figs. 4.8a and 4.10b). They can use imagery to anticipate how to turn a shape in order to fit it in a shape frame; so there is less trial and error when physically decomposing a shape.

Example 4.3

For the frame task shown in figure 4.8a, Jenna quickly decided on the correct pattern blocks for each shape, turning each shape in her hand as she moved to correctly place it on the matching shape outline; she did not use trial and error.

However, students who are reasoning at level 2 still cannot create images that are accurate and stable enough to visualize more complex shape embeddings. For instance, students reasoning at level 2 generally find it difficult to imagine or draw where the dark blue rhombuses fit in figure 4.10a. Although students can fairly easily solve these problems physically, mental visualization is the difficulty.

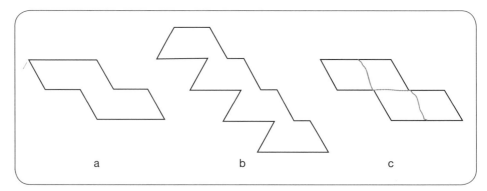

Fig. 4.10

At level 2 students can also begin to use physical pattern blocks to correctly reason about which shapes have greater area as in the following task:

Use pattern blocks to decide which shape (fig. 11a) has more area or room inside it, or if the two shapes have the same area or room inside.

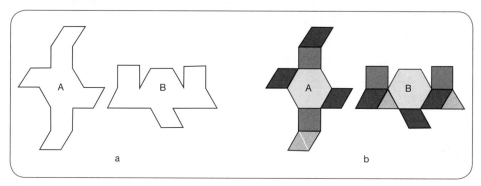

Fig. 4.11

Jack: [*Makes shape A with the set of pattern blocks shown in figure 4.11b, then uses those same pattern blocks to make shape B.*] Because I used the same blocks to make the shapes (fig. 4.11b), they have the same room inside.

Instruction to Help Students Move from Level 1 to Level 2

Tasks like Types 1–4 in Appendix C (page 145) should be given in order. After a task is given, ask the students to place the pattern blocks they think will work in a tray: "Which pattern blocks do you think will make this shape? Place the blocks in the tray." Then have students check their prediction with blocks, exchanging blocks if needed. The goal is for students to *visualize* which shapes match before testing their visualizations with actual pattern blocks.

Using tasks like Types 5 and 6 in order, give students task sheets and pattern blocks, and ask: "Try to make the shape with the blocks. Are there other pattern blocks that will make the shape?" The goal is for students to use physical trial and error to start decomposing shapes in which not all the component shapes are clearly distinguishable (e.g., the shape in fig. 4.10b is a Type 5 task). Students should become successful with Type 5 tasks, in which many shape sides are explicit, before moving to Type 6 tasks, in which few sides of the shapes are shown.

Level 3

The student accurately visualizes matching shapes and visualizes decomposing by drawing. Students can create images that are accurate and stable enough to visualize more complex shape decompositions. For instance, students reasoning at level 3 can draw where the dark blue rhombuses fit in figure 4.10a (see drawing in fig. 4.10c). They do not need to physically place dark blue rhombuses to make an accurate drawing.

Instruction to Help Students Move from Level 2 to Level 3

Using tasks like Types 5 and 6, we have students *draw* where they think blocks will fit: "Which pattern blocks do you think will make this shape? Draw where you think they will go." Students check their drawings with blocks. If students' drawings are incorrect, we should have them remove the blocks and try again to draw where the blocks should be placed. The goal is for students to strengthen their imagery so that it is accurate even during complex tasks in which several shapes must be thought of at the same time.

Altering Instructional Tasks to Support Student Sense Making

If students have difficulty with frame tasks in which shapes are deeply embedded, sometimes using color frames can help them. For instance, for the frame tasks below, we can ask, "What pattern blocks do you think were used to make this shape?" Or, "Can you make this shape with rhombuses?" Or, "This shape (fig. 4.12a) can be made with five dark blue rhombuses. Figure out how."

Fig. 4.12

Quick-Image Activities: Building Mental Visualization[8]

Another powerful type of instructional activity for building students' geometric visualization uses "quick-image" tasks. To implement these tasks, we give each student a small set of pattern blocks. We tell students that we are going to show a design made from pattern blocks on an overhead projector or document camera for three seconds and then we will hide it. They are to make the design with their pattern blocks (see examples in fig. 4.13). We turn on the overhead projector/ document camera for three seconds, show the design, then turn the machine off. We give students, working individually, a few minutes to try to create the design with their blocks. Then we show the design for another three seconds and give students several more minutes to think about the problem. Finally, we turn on the display and ask students if the designs they created are correct. After students have completed their final checks of their designs, we have them explain the strategies they used to figure out how to create the design we showed them, and how they remembered what the design looked like. We can repeat this procedure with several pattern-block designs, using designs that are appropriate for the students' current levels of reasoning. We start with simple designs that consist of just two pattern blocks, then increase the number of pattern blocks as students' capability increases. (Before doing quick image activities, it is important to be sure that, given unlimited time, students can duplicate a design shown to them.)

While viewing students' work, we should consider the following questions: How accurate are students' first predictions; second predictions? Do students' predictions use the correct shapes? Do students' predictions put correct shapes in the correct locations?

Fig. 4.13. Using composite units to structure pattern-block designs

Once students have developed proficiency in visualizing pattern-block shape decompositions, instruction should target developing their ability to decompose shapes into *spatial composites*. One way to encourage and support students' use of composite units is with "predict-and-cover" tasks. After looking at these tasks, we examine levels of sophistication in students' reasoning about the tasks and instructional activities that promote students' development of this critically important mental process.

Predict-and-Cover Tasks: Spatial Structuring by Iterating Composite Units[9]

Students form a composite unit by mentally uniting a collection of objects, then iterating that unit by repeating it. For instance, if a student thinks of combining two light blue triangles to form a dark blue rhombus and iterates this unit to cover a larger shape with copies of this pair-of-triangles-as-rhombus (fig. 4.14), then the student has iterated the composite unit of two triangles.

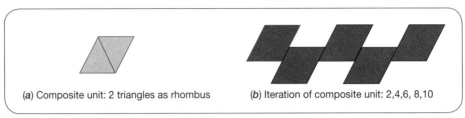

(a) Composite unit: 2 triangles as rhombus   (b) Iteration of composite unit: 2,4,6, 8,10

Fig. 4.14

Predict-and-cover tasks present sequences of three problems that encourage students to spatially structure shapes by iterating composite units (fig. 4.15; Battista 2012b), an essential mental process in geometric measurement. The prediction step in these tasks is critical for imagery and mental-model building. When asked to predict an answer, students must reflect on the situation and create a mental model to visualize possible solutions. As they check their solutions, they revise their mental models, making them more accurate. These tasks involve students in making sense (SMP 1g, 2a) and in constructing and critiquing ideas as they explain their enumeration strategies (SMP 3).[10]

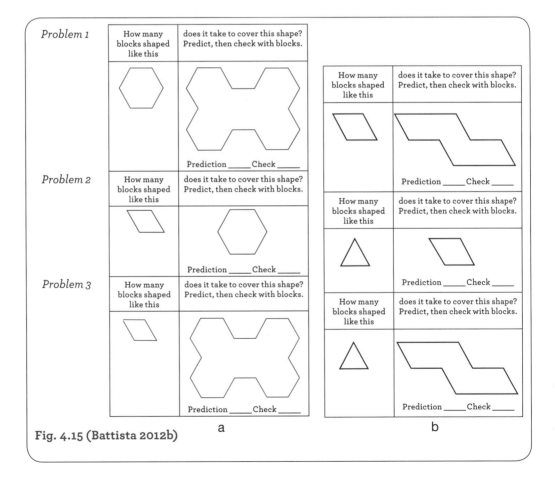

Fig. 4.15 (Battista 2012b)

Levels of Sophistication in Students' Reasoning About Pattern-Block Composite Units

Level 0

The student makes no explicit attempt to use composite units.

Example 4.4

Teacher:	[Problem 1, fig. 4.15a] How many hexagons do you predict it takes to cover this shape?
Max:	5. [*Max checks and finds that 5 is correct.*]
Teacher:	[Problem 2, fig. 4.15a] How many dark blue rhombuses do you predict it takes to cover this shape [hexagon]?
Max:	4. [*Max checks and finds that 3 is the correct answer.*]
Teacher:	[Problem 3, fig. 4.15a] How many dark blue rhombuses do you predict it takes to cover this shape?
Max:	[*Covering the five-hexagon shape with rhombuses as shown in fig. 4.16*] I was thinking that it would probably be 14.

As shown in figure 4.16a, Max did not use three-rhombus composites (fig. 4.16b) to enumerate rhombuses in the five-hexagon shape. Because he did not structure the large shape as 5 composites of 3 rhombuses, Max's structuring of the large shape into rhombuses was not yet powerful enough to correctly enumerate the rhombuses. Note, however, that Max did create an interesting repeating structure of composites of 2 rhombuses.

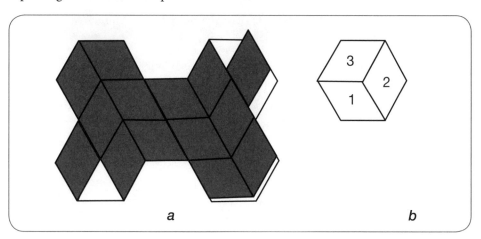

Fig. 4.16

Level 1

The student makes an explicit attempt to use composite units, but because of heavy imagery demands, makes visualization errors.

Example 4.5

Branford: [*Working on fig. 4.15b*] It takes 2 of these to make 1 of these. [*He makes a dark blue rhombus out of 2 light blue triangles.*] It takes 4 of these [rhombuses] to make this. [*He uses rhombuses to make the four-rhombus target shape.*]

Branford: [*To determine how many triangles it takes to make the larger shape, Branford moves a pair of fingers inside the shape (fig. 4.17), skip counting by 2 as he points to the positions.*] 2, 4, 6, 8, 10. Maybe 10.

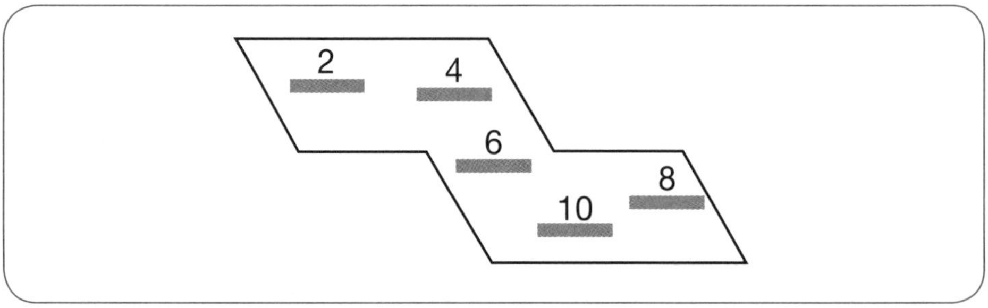

Fig. 4.17

Teacher: Okay, let's check your answer with light blue triangles. . . .

Branford: [*After correctly placing triangles on the target shape*] 2, 4, 6, 8. Oh, it's only 8.

Although Branford had just found, using pattern blocks, that 4 rhombuses cover the large shape (Problem 1), and that 2 triangles make a rhombus (Problem 2), on Problem 3, he incorrectly decomposed the large shape into 5 rhombuses instead of 4. Branford was unsuccessful because he could not hold in mind an image of where the 4 embedded rhombuses could be placed *at the same times as* holding in mind how 2 triangles make each rhombus. Branford could, however, easily check his answer with physical triangles, something that students at Level 0 have difficulty with.

Example 4.6

Like Branford, Adam correctly answered Problems 1 and 2 for the task shown in figure 4.18, determining physically that 4 hexagons fit in the target shape and 3 rhombuses fit in the hexagon. However, when he tried to use composite units to determine how many rhombuses fit into the target shape, he could not maintain a correct image of how many rhombuses fit into the hexagons, counting 2 instead of 3.

Adam: [*Drawing the hexagons shown in fig. 4.18*] Then it would make maybe 4 [hexagons]. [*Counting each hexagon as 2 rhombuses*] 2, 4, 6, 8.

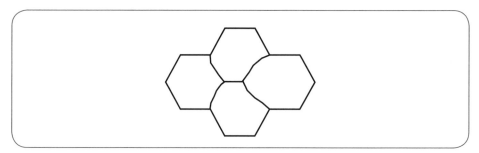

Fig. 4.18

Instruction for Moving Students from Level 0 to Level 1

To encourage students to use composite units, we might continue with Max like his teacher did for the task shown in figure 4.15a. As shown below, this intervention helped Max use composite units of 3 rhombuses as a hexagon, moving him to level 1, but he was only partially successful in this reasoning.

Teacher: At one point, you said that 3 of these made one of these [*pointing to a rhombus then a hexagon*], then you said that there would be 5 of these [*pointing to a hexagon*] here, here, here, here, and here [*pointing to the 5 embedded hexagons*]. What do you think about using that idea?

Max: Because it would equal 3 of these [*showing one rhombus on top of a hexagon*]. It would go 3, 6, 9, 12 [*pointing to the 4 outer hexagons embedded in the 5-hexagon shape*], and I thought that. But it was 2 more over it.

Teacher: Where did those extra 2 go?

Max: Probably over here in the middle somewhere [*pointing*].

To help students move to level 1, we can also use continue-the-pattern problems to encourage them to start seeing structure and iterating composite units. The pattern in figure 4.19 can be created by iterating a composite unit that consists of 2 triangles and a rhombus. Of course, some students might structure the pattern as an alternating repetition of 2 stacked triangles and 1 dark blue rhombus.

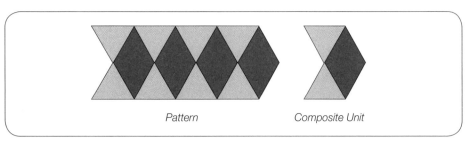

Pattern   Composite Unit

Fig. 4.19

Level 2

Students correctly visualize iterating composite units.

Students explicitly and correctly structure a larger shape into iterations of composite units of smaller shapes. They coordinate iterations of spatial composites with appropriate numerical reasoning (e.g., skip counting by 3 as they iterate a spatial composite of 3 rhombuses).

Example 4.7

Bob is trying to visualize and enumerate dark blue rhombuses in figure 4.20a.

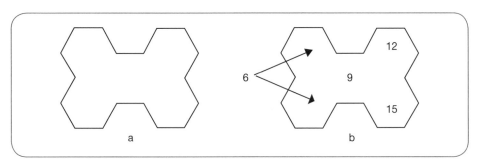

Fig. 4.20

Bob: It takes 3 [rhombuses] to cover one of these [*pointing to the upper left hexagon; he had already discovered this*]. [*Pointing to the 2 leftmost hexagons in the shape in fig. 4.20b*] 6; [*pointing to the middle*] this will take 3, so that's 9; [*pointing to the upper right*] 12; [*pointing to the lower right*] 15.

Example 4.8

Michelle is trying to enumerate light blue triangles in figure 4.21. She has already used blocks to verify that 2 triangles make a rhombus and that 5 rhombuses cover the shape in fig. 4.21.

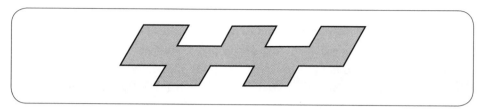

Fig. 4.21

Michelle: [*Looking at fig. 4.21, she taps each finger twice on the table*] 1, 2 [*first finger*], 3, 4 [*second finger*], 5, 6 [*third finger*], 7, 8 [*fourth finger*], 9, 10 [*fifth finger*]. There's 10.

Michelle replaced finger pointing at individual spatial units (in this case triangles) with a slightly more abstract procedure of counting her tapping gestures.

Instruction for Moving Students from Level 1 to Level 2

When students are predicting how many of the small blocks it takes to cover the target shape, we can leave their covering of the large block with small blocks as well as their covering of the target shape with large blocks visible (fig. 4.22) . We can ask: "Can knowing about what you did here and here help you?"

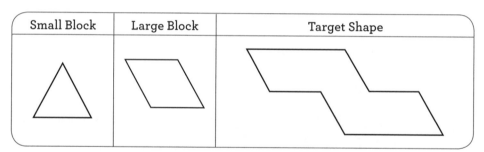

Small Block	Large Block	Target Shape

Fig. 4.22

As students work on the predict-and-cover tasks in Appendix D (page 149), if students have difficulty, we can have them first *draw* how the larger pattern-block covers the target shape and then ask them to solve the problem. If they still have difficulty, we can have them draw how the smaller pattern-block fits in one of the larger pattern blocks, then predict how many of the smaller pattern blocks fit in the target shape. Of course, students should check their drawings and answers with pattern blocks. Also, after students have covered about half of a shape frame with small blocks, often it is useful to ask how many they would predict altogether. Because they have already covered half the shape, students are often able to make better predictions. (Further discussion of how to help students in Level 2 can be found in the section on instructional helping strategies on page 125.)

Level 3

Students correctly perform numerical iteration without physical materials.
Students abstract the spatial structure of composite-unit coverings sufficiently
to enable them to represent the situation strictly numerically. For instance, in a
classroom discussion, students are reporting on their investigation of the shape
in figure 4.23. They have established that it takes 4 hexagons to cover the shape.

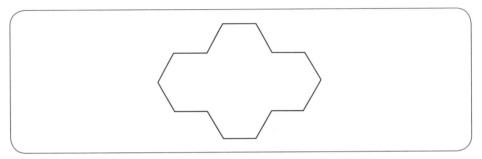

Fig. 4.23

Teacher:	Now that you know that this shape contains 4 hexagons, how many trapezoids does it have?
David:	8, because I know 2 trapezoids make a hexagon, so I doubled the number.
Teacher:	How about light blue triangles?
Keira:	24.
Teacher:	Why?
Keira:	Because 6 light blue triangles make a gray hexagon, and 6 and 6 is 12, and 12 and 12 is 24.
Lenny:	I added up 8 threes because there are 8 trapezoids, and 3 triangles make a trapezoid.
Teacher:	How about dark blue rhombuses?
Jeri:	3 dark blues make a hexagon, and 3 times 4 is 12.

The solutions of these four students are more sophisticated than those in
level 2 because these students did not need to look at or point to shapes within
the large shape to carry out their enumerations. They abstracted the situation
away from its spatial context and transformed it into a numeric representation.
Also, they used more sophisticated numerical operations of addition and
multiplication instead of counting by ones or threes. Reasoning at level 3 uses
mathematical practices of abstraction (SMP 2a) and modeling (SMP 4a). Note
also how students' justifications for answers to predict-and-check problems

evolve from physical demonstrations at level 2 to arithmetic arguments at level 3 (SMP3).

Instruction for Moving Students from Level 2 to Level 3

Predict-and-cover tasks can be presented on an overhead projector so students can see the shapes but not manipulate them. For instance, for the first predict-and-cover task in Appendix D, we ask students each of the first two questions on the student sheet and have a class discussion about any answer disagreements until the class establishes the correct answers to these questions. (If need be, we can illustrate the answers on the overhead.) Next, we have students answer the third question. The goal is for students to answer the third question as students in level 3 do: "There are 5 dark blue rhombuses in the target shape. There are two light blue triangles in each rhombus. So there are 5 times 2 equals 10 light blue triangles in the target shape."

Structuring Rectangles as Composites of Squares: Levels of Sophistication in Students' Spatial Structuring of Rectangular Arrays

A core idea in developing reasoning and sense making to determine areas of rectangles is structuring and meaningfully enumerating arrays of squares (see Battista et al. 1998; Battista 1999, 2012b).

Level 0

The student structures unit-square counting as a path with no array structure and no use of composite units. This level of reasoning is exemplified by Katy. Her counting paths meandered about so much that she could not keep track of where the paths had been or where they should go next.

Level 1

The student structures part of an array as composite units (Example 4.9), or the student tries to mentally structure the whole array as composite units, but because of heavy imagery demands, makes visualization errors (Example 4.10).

Example 4.9
Bill was asked to predict the number of squares that cover the rectangle shown in figure 4.24a.

> Bill: First I count the bottom, and there's 6. [*Moving his hands inward as shown in fig. 4.24b*] So the top and bottom would equal 12. And these 2 [*pointing to the middle squares on the right and left sides*] would be 14. [*Using fingers to estimate where individual squares were located*] I'd say maybe 12 in the middle; 12 + 12 = 24. So I'd say 24.

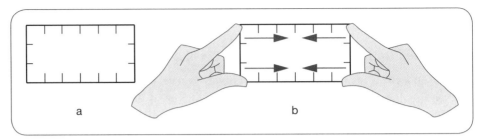

Fig. 4.24. Bill's work

Bill's first inference, that the top has the same number of squares as the bottom, is the first step that many students take toward using composite units in a way that can eventually lead to a row-by-column structuring of a rectangular array of squares. However, because Bill could not apply this row structuring throughout the whole rectangle, he was unable to correctly locate squares in its middle section.

Example 4.10
Joe was shown that five squares fit across the top of a rectangle and that seven squares fit down the middle (then the squares were removed).

Joe:	5, 10, 15, . . . , 45 [*motioning across rows inside the rectangle*].
Teacher:	How did you get that?
Joe:	I was trying to guess where the bottoms of the squares were. [*Joe places 7 squares down the right column, then quickly points to each and says, "35"*] 5, 10, 15, . . . , 35; 5 times 7. I'm positive.

Joe structured the array into rows. However, before he checked his answer by placing squares down the middle of the rectangle, he incorrectly visualized how the rows could be iterated to cover the rectangle. He needed an actual column of squares to correctly iterate the seven rows of five squares.

Level 2

The student correctly visualizes iterating composite units and coordinates spatial and numerical iteration. At this level, students have abstracted the row-by-column structure of an array to a degree that enables them to enumerate rows of squares using their fingers, oral skip counting, or paper and pencil. They do not need actual squares or drawings to enumerate the squares.

Example 4.11
Paul was given the same problem as Joe: he was shown that five squares fit across the top of a rectangle and that seven squares fit down the middle (then the squares were removed).

Paul:	You said 7 up, right? Five across, 7 down; (pause) 7 down, 5 across. [*Motioning across the top 3 rows of the rectangle*] 5, 10, 15. [*Counting on 7 fingers*] 5, 10, 15, 20, 25, 30, 35; 35.
Teacher:	How did you know to stop at 35?
Paul:	Because when I lifted up my last finger I knew that was 7.
Teacher:	What did that tell you? Why 7?
Paul:	There are only 7 down that way [*motioning vertically down the middle of the rectangle*].

Level 3

The student correctly enumerates composite units without physical-spatial material and can explain/justify why enumeration procedures work.

Example 4.12

A rectangle measures 8 inches by 7 inches. What is its area?

Grace:	It's 8 times 7 equals 56.
Teacher:	How do you know? Why does that procedure work?
Grace:	It's 8 by 7 inches. So it's like 7 rows of 8 squares, 7 groups of 8, which is 7 times 8 or 56 squares.

Spatial Sense Making Under the Microscope: Coordinating Rows and Columns (from Battista 1999)

As we have seen, constructing a proper structuring of two-dimensional arrays of squares is quite difficult for many students. The following example illustrates how second grader Amy made spatial and numerical sense of an array by constructing a row-by-column structuring for it. She is predicting how many squares will cover a 4-inch-by-3-inch rectangle, having been shown only that 4 squares fit across its top and 3 down its left side. Note how Amy restructures her mental model of the array as she draws where she thinks the squares go and reflects on her drawing. [If Amy had not drawn squares unprompted by the teacher, it would have been very useful for the teacher to suggest doing so. In earlier problems, Amy had been asked to draw squares, so she understood that drawing was a useful strategy available to her.]

Amy:	[*Pointing as shown in fig. 4.25a*] There's 4 here [*top row*] and 4 here [*bottom row*] plus 2 here [*one on the left side, one on the right*], and that equals 10. [*Pointing to the middle of the rectangle interior*] But I'm not sure if there's 2 in the middle or 1 in the middle.
Amy:	[*After drawing perimeter squares but leaving the middle blank (fig. 4.25b)*] I think there's 2 in the middle or 1 in the middle.

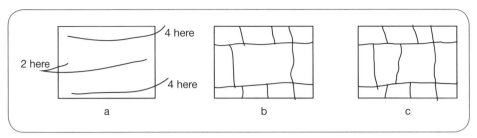

Fig. 4.25 (Battista 1999)

Teacher: Where are the 2 in the middle?

Amy draws the middle vertical segment in the second row as in figure 4.25c.

Teacher: So how many do you have in there now?

Amy: So 12 or 11. I think there's 2 in the middle or 1 in the middle. [*Motioning across the second row*] I just changed my mind. I think there's 4 in the middle [*pointing to the second row of her drawing*]. . . . There's going to be 2 in the middle because if there's only 3 here [*motioning to the right column*] and 3 here [*motioning to the left column*], then going across would be 4 [*motioning across the second row*]; 12.

Teacher: Why do you now think 12?

Amy: Because there's 4 on the top [*motioning to the top row*], so going across down here [*motioning to the second row*] would be the same as going across up here [*motioning to the top row*], so there's probably 2 in the middle. Altogether there are 4 [*pointing to the 4 in the middle row*].

At first, Amy saw the set of squares that covered the rectangle as a set of disjointed parts: top row, bottom row, separate side squares, and an indeterminate center. She was able to see the center of the rectangle as two squares, and thus its middle row as four, only when she considered the two outer squares in the middle row (fig. 4.25b) not as separate but as members of the right and left columns. As she coordinated these columns with squares in the top row, she was able to see the equivalence of the first and second rows. This coordinating action enabled Amy to make sense of the array by giving it a row-by-column structuring.

Instructional Activities with Rectangles (from Battista 1999)

To develop the ability to properly structure two-dimensional arrays of squares, students need numerous opportunities to structure such arrays and to reflect on the appropriateness of their structurings. A useful way of presenting such opportunities is to utilize problems similar to those previously illustrated and as shown in figure 4.26 (see Battista et al. 1998; Battista 1999, 2012b). For instance, before giving them out to the students, enlarge each rectangle in the figure so its dimensions are in inches. Once students have a rectangle, show them how one plastic inch square fits on one square indicated in the rectangle. Students first predict how many squares it takes to cover the rectangle, then check their predictions with plastic squares.

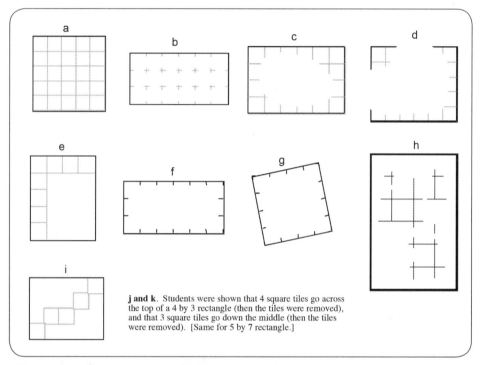

j and k. Students were shown that 4 square tiles go across the top of a 4 by 3 rectangle (then the tiles were removed), and that 3 square tiles go down the middle (then the tiles were removed). [Same for 5 by 7 rectangle.]

Fig. 4.26

As a variation of this activity, after students have made their first prediction, we can have them draw how they think squares will cover the rectangle, make another prediction, then check their predictions with plastic squares. Many students will be able to make a correct prediction after drawing squares on a rectangle, but their structuring will not be strong enough to make a correct original, strictly mental, prediction. If students need to put squares on the rectangles to get a correct count, after they do so, we should have them remove the squares, then try to draw where the squares were, again checking their answers by putting squares on the rectangles.

It's best to start with rectangles that give the most graphic information about the location of squares then gradually move to rectangles that give less information (the rectangles in fig. 4.26 are listed roughly in order of difficulty). We should give students several problems of each type of graphic representation so that they have an opportunity to develop adequate structuring for that type before moving on to more difficult problems.

Connecting Pattern-Block Reasoning to Reasoning About Area with Unit Squares

The task shown in figure 4.27 helps students connect the non-measurement reasoning they used in the frame tasks with measurement (number-based) reasoning used in the pattern-block design and array tasks.

Which shape has more area or room inside, or do they have the same amount of room? (Students are provided with the two shapes drawn on inch dot paper (fig. 4.27), along with plastic inch squares.)

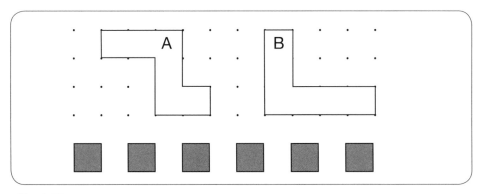

Fig. 4.27

Mario: I covered the inside of shape A with squares, then moved these squares to cover shape B. So I think they have the same room inside because the same squares fit in them. [*This is the same type of reasoning used on fig. 4.11.*]

Maria: I counted 6 squares that fit in shape A and 6 squares that fit in shape B. So they are the same. [*If no student counts, you might ask the class "Could counting anything help you answer this question?"*]

It is important for students to see and relate the non-measurement, movement-based reasoning of Mario to the measurement, count-based reasoning of Maria (SMP 1b, 1g, 2a, 4c, d, 6b; PS 4a, 5b). Problems such as these help students discover and *make sense of* the idea that areas can be compared by counting unit squares. *Counting works because if the number of squares that cover each shape is the same, then the squares that cover one shape exactly can be moved to cover the other shape exactly; that is, one shape can be decomposed to make the other shape exactly.* Understanding this relationship between non-measurement and measurement reasoning is fundamental to using numbers to represent length, area, and volume measurements (SMP 1, 1b, 1g, 2a, 2d, 3d, 4; PS 4; PS 5).

Teacher Dialogue to Support Student Reasoning[11]

We now examine an extended example of a teacher-student dialogue that illustrates how teachers can help students progress to higher levels of reasoning about shape decomposition. Such dialogue is critical in classrooms that support student sense making. This description also provides an in-depth look at the complexity of a student's struggle to make sense of and adopt more sophisticated reasoning. The teacher's dialogue with Carla helps Carla not only to solve the problem she is working on but also to use more sophisticated reasoning to solve the problem. The teacher's instructional dialogue targets both Carla's use of imagery and her use of composite units.

> *How many blocks shaped like this [hexagon pattern block] does it take to cover this shape (fig. 4.28)?*

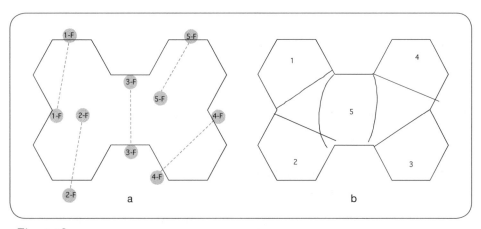

Fig. 4.28

Carla:	[*Carla uses 2 fingers and moves them around the shape to make her prediction (see locations of Carla's finger-pair locations in fig. 4. 28a).*] 1, 2, 3, 4, 5.
Teacher:	Why do you think 5?
Carla:	Because it would go 1 [*points and draws a line segment (fig. 4.28b)*], 2 [*points and draws a segment*], 3 [*points and draws a segment*], 4 [*points and draws a segment*], and then that would be 5 [*draws two arcs around the middle*].
Teacher:	Okay! Go ahead and check with blocks.
Carla:	[*Correctly placing hexagons on the shape and counting (fig. 4.29a)*] 1, 2, 3, 4, 5. 5!

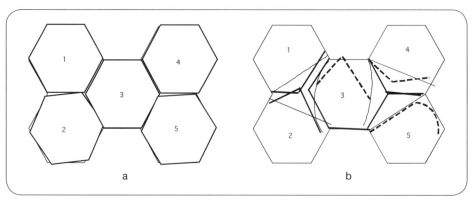

Fig. 4.29

Although Carla's prediction was correct, the way she originally drew and pointed to hexagon locations suggest that her imagery was approximate and inexact. She was reasoning at level 2 about decomposing shapes. Also, note the role of different *representations* in Carla's sense making for the decomposition of shapes (PS 5). Her mental imagery was approximate (fig. 4.28a), her drawings were better but still approximate (fig. 4.28b), and her placement of physical pattern blocks was completely accurate (fig. 4.29a). The teacher next gave Carla an opportunity to improve her imagery and drawings.

The teacher then asked Carla to take one block away at a time and draw the lines. The dark and dotted line segments in figure 4.29b show how Carla drew the sides of the hexagons as she removed hexagon pattern blocks one at a time. The dotted segments indicate where she drew her first attempts, but the teacher had her try again because they were inaccurate. Again there is some indication that Carla's imagery is approximate.

Teacher: All right! How many blocks like this [the rhombus on the predict-and-cover student sheet, fig. 4.15a] does it take to cover this shape [the hexagon on the same student sheet]?

Carla: I think probably 3.

Teacher: Can you show me where you got 3?

Carla: [*Drawing line segments and counting as shown in fig. 4.30a*] Um, one would probably go right here, one would probably go right here, and one would probably go right here.

Teacher: All right! Go ahead and check.

Carla puts the dark blue rhombuses on the hexagon (fig. 4.30b).

Teacher: So, how many did you get?

Carla: 3!

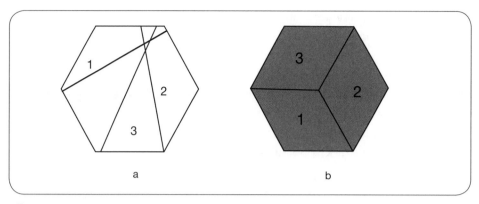

Fig. 4.30

Again, even though Carla's numerical prediction was correct, she seemed to be struggling to correctly visualize and structure the 3 rhombuses that cover the hexagon. She was still reasoning at level 2 about decomposing shapes.

Teacher: Okay! How many blocks shaped like this [*pointing to the dark blue rhombus*] does it take to cover this shape [*pointing to the five-hexagon shape at the bottom of fig. 4.15a* (Battista 2012b)]?

Carla: [*Counting as she draws where she thinks rhombuses fit on the 5-hexagon shape*] 1, 2, 3, 4, 5, 6, 7, 8, 9, 10, 11, 12, 13 (fig. 4.31a).

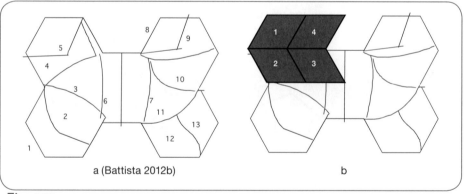

Fig. 4.31

Teacher: So, you got 13? [*Carla nods her head yes.*] Go ahead and check. [*Carla puts the first 4 dark blue rhombuses on the shape and then counts them (fig. 4.31b).*] 1, 2, 3, 4.

Note how the teacher engaged Carla in discussion about Carla's mental models for covering the various shapes. This discussion reveals that Carla either made no explicit attempt or was unable to utilize the hexagon as a composite unit of rhombuses. Although Carla correctly predicted the number of hexagons that fit in the five-hexagon target shape and the number of rhombuses that fit in the hexagon, her attempts to draw and count the number of rhombuses that fit in the five-hexagon target shape indicate that her mental visualization was too inaccurate for decomposing the five-hexagon shape into rhombuses. Furthermore, her imagery difficulties even extended to physically covering the five-hexagon shape with rhombus pattern blocks (fig. 4.31b). Carla was at level 0 for reasoning about pattern-block composite units and level 1 in her reasoning about decomposing shapes. The following description illustrates how the teacher helped Carla:

Teacher: [*Takes a single hexagon and puts it on the table next to the 5-hexagon shape.*] Now this shape fits in here [*top right portion of the five-hexagon shape*], right? [*Carla nods her head yes; the teacher takes the hexagon off the 5-hexagon shape.*]

Teacher: Why don't you take these [*pointing to the dark blue rhombuses*] and build on top of that [*the removed hexagon*] to see what it looks like?

Carla puts 3 dark blue rhombus pattern blocks on the hexagon pattern block. She takes them off and puts them on the top right hexagon of the five-hexagon shape. She builds on the hexagon again and places these 3 rhombuses on the bottom right hexagon of the five-hexagon shape. She puts 3 more rhombuses on the hexagon and puts them on the bottom left hexagon of the five-hexagon shape, then does the same for the upper left hexagon of the five-hexagon shape (fig. 4.32). Carla then counts the rhombuses she has on the frame. She starts over with 1 and 2.

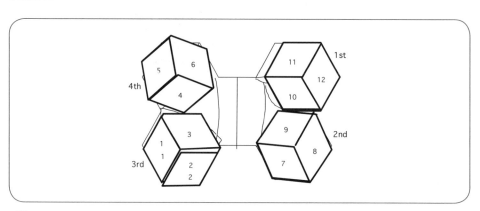

Fig. 4.32

Teacher: How many do you have so far?

Carla: 12!

Teacher: 12 so far. Let's look at this. Is this on there right [*the top left hexagon*]? I see a little bit out here. Can you turn that?

Carla: [*Carla looks at the top left rhombuses and turns them to fit the shape.*] There!

Teacher: Good. I think you've got it.

Carla puts a rhombus (13) in the middle section of the five-hexagon shape (fig. 4.33a).

Teacher: You're not going to put it on here [*pointing to the hexagon*]?

Carla: Yeah, I guess so.

Carla takes rhombus 13 off the shape and puts it on the hexagon pattern block along with 2 other rhombuses. She then puts them on the five-hexagon shape as shown in fig. 4.33b.

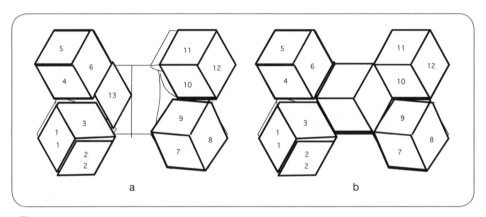

Fig. 4.33

Teacher: You did it Carla, didn't you? [*Carla nods her head yes.*]

Teacher: Okay! Now we have to check. Count for me out loud so I know which ones you're pointing to.

Carla: [*Correctly counting rhombuses*] 1, 2, 3, 4, 5, 6, 7, 8, 9, 10, 11, 12, 13, 14, 15.

In this episode, the teacher found an ingenious way to help Carla correctly iterate and enumerate spatial composites of 3 dark blue rhombuses that cover the large shape. The teacher had Carla *physically create each composite* separately, off the shape, then place them onto the five-hexagon shape. Repeated physical experiences like this can help students improve their visualization and their use of composite units. Note that covering the target shape with rhombuses is much easier to visualize if one uses composite units; indeed, in general, structuring a shape as iterations of composite units can not only help in making sense of the covering but it can also vastly improve the accuracy of visualizations. That is why row-column structuring is so powerful for area measurement.

The next instructional move to engender further progress for Carla would be to give her Predict-and-Check Task 2 (Appendix D, page 150) to determine if she has abstracted the three-rhombuses-as-hexagon composite sufficiently to solve this new task without physical materials. If she cannot, we would let her solve it physically, as the teacher did in the previous episode, then give her another task that involves the three-rhombuses-as-hexagon composite. Many students need repeated uses of physical materials, along with reflection on what they are doing, before they are able to construct mental models that are stable enough to visualize the situations in this type of problem solving.[12]

Conclusion

Note how the three sets of levels of sophistication are related. First, students build sufficient imagery for shape matching and decomposition. Then they extend this imagery to iterating spatial composites. Finally, they abstract their spatial composite-unit reasoning sufficiently to systematically enumerate composite units and extend this reasoning to finding areas of rectangles in a meaningful way.

Designing and implementing instruction that supports students' meaningful learning about geometric decomposition and structuring must be based on firm understanding of the development of students' reasoning about these concepts, that is, knowledge of learning progressions. Such understanding is essential to teaching in a way that is consistent with professional recommendations and modern research on students' mathematics learning. To enable students to achieve more than superficial understanding of shapes and their decomposition, instruction must focus on, guide, and support students' movement through the increasingly sophisticated levels of reasoning and sense making described in this chapter.

Endnotes

1. Much of the research and development referenced in this chapter was supported in part by the National Science Foundation under Grant Numbers 0099047, 0352898, 554470, 838137, and 1119034. The opinions, findings, conclusions, and recommendations, however, are mine and do not necessarily reflect the views of the National Science Foundation.

2. These are sobering findings, given that for many students in these grades, traditional instruction uses rectangular arrays as one model to *give* meaning to multiplicaton, assuming that students see such arrays as sets of equivalent columns and rows.

3. Descriptions of the development of other aspects of geometric reasoning appear in numerous other publications (Battista 2003, 2007ab, 2009, 2012b; Clements and Sarama 2009; Sarama and Clements 2009). Much more extensive treatments of students' reasoning about arrays of squares and area can be found in Battista (1999, 2012b) and Battista et al. (1998).

4. A numbered/lettered summary of the CCSM Mathematical Practices and NCTM Process Standards is given in the appendix to this book.

5. It is assumed that all students pass through almost all of the levels in learning progressions. What varies is the speed at which they pass through the levels and the amount of instructional scaffolding students need to pass through the levels.

6. A web applet is available from the author on request.

7. See Akers et al. 1997; Battista 2012; Clements and Sarama 2009; Sarama and Clements 2009.

8. See Wheatley (2007) for similar activities.

9. I developed this type of task while working on the *Investigations in Number, Data, and Space* curriculum. See Akers et al. (1997). See also Battista (2012a).

10. If students have difficulty counting the number of blocks that actually cover a shape, have them count the blocks as they remove them one at a time.

11. For additional instructional suggestions, see Battista (1999, 2012a).

12. Note that in most pattern block sets, the hexagon is yellow, the trapezoid is red, the larger rhombus is blue, and the triangles are green. We changed the colors to work in the two-color format of this chapter.

References

Akers, J., M. T. Battista, A. Goodrow, D. H. Clements, and J.Sarama. *Shapes, Halves, and Symmetry.* Palo Alto, Calif.: Dale Seymour Publications, 1997.

Battista, M. T. "The Importance of Spatial Structuring in Geometric Reasoning." *Teaching Children Mathematics* 6, no. 3 (November 1999): 170–77.

Battista, M. T. *Shape Makers: Developing Geometric Reasoning in the Middle School with the Geometer's Sketchpad.* Berkeley, Calif.: Key Curriculum Press, 2003.

Battista, M. T. "Understanding Students' Thinking about Area and Volume Measurement." In *2003 Yearbook, Learning and Teaching Measurement,* edited by D. H. Clements, pp. 123–124. Reston, Va.: National Council of Teachers of Mathematics (NCTM), 2003.

Battista, M. T. "The Development of Geometric and Spatial Thinking." In Second Handbook of Research on Mathematics Teaching and Learning, edited by F. Lester, pp. 843–XXX. Reston, Va.: National Council of Teachers of Mathematics (NCTM), 2007a.

Battista, M. T. "Learning with Understanding: Principles and Processes in the Construction of Geometric Ideas." In 69th NCTM Yearbook, The Learning of Mathematics, edited by M. E. Strutchens and W. G. Martin, pp. 65–69. Reston, Va.: NCTM, 2007b.

Battista, M. T. "Highlights of Research on Learning School Geometry." In *2009 Yearbook, Understanding Geometry for a Changing World,* edited by T. Craine and R. Rubenstein, pp. XX–XX. Reston, Va.: NCTM, 2009.

Battista, M. T. "Fifth Graders' Enumeration of Cubes in 3D Arrays: Conceptual Progress in an Inquiry-Based Classroom." *Journal for Research in Mathematics Education* 30, no. 4 (July 1999): 417–48.

Battista, M. T. (2011). "Conceptualizations and Issues Related to Learning Progressions, Learning Trajectories, and Levels of Sophistication." *The Mathematics Enthusiast* 8(3). Article 5. Available at: http://scholarworks.umt.edu/tme/vol8/iss3/5 pp507-570

Battista, M. T., and M. Berle-Carman. *Containers and Cubes.* Palo Alto, Calif: Dale Seymour Publications, 1996.

Battista, M. T., and D. H. Clements. "Students' Understanding of Three-Dimensional Rectangular Arrays of Cubes." *Journal for Research in Mathematics Education* 27 (May 1996): 258–92.

Battista, Michael T. (2012a). *Cognition Based Assessment and Teaching of Geometric Measurement (Length, Area, and Volume): Building on Students' Reasoning.* Portsmouth, N.H.: Heinemann, 2012a.

Battista, Michael T. (2012b). *Cognition Based Assessment and Teaching of Geometric Shapes: Building on Students' Reasoning.* Portsmouth, N.H.: Heinemann, 2012b.

Battista, Michael T., Douglas H. Clements, Judy Arnoff, Kathryn Battista, and Caroline V. A. Borrow. "Students' Spatial Structuring and Enumeration of 2D Arrays of Squares." *Journal for Research in Mathematics Education* 29 (Nov. 1998): 503–32.

Ben-Chaim, D., G. Lappan, and R. T. Houang. (1985). "Visualizing Rectangular Solids Made of Small Cubes: Analyzing and Effecting Students' Performance." *Educational Studies in Mathematics* 16 (1985): 389–409.

Clements, D. H. and J. Sarama. *Learning and Teaching Early Math: The Learning Trajectories Approach.* New York: Routledge, 2009.

Gelman, S. A. and C. W. Kalisch. "Conceptual Development." In *Handbook of Child Psychology.* Vol. 2 of *Cognition, Perception, and Language,* edited by Deanna Kuhn and Robert Siegler, pp. 687–733. Hoboken, N.J.: Wiley, 2006.

Hirstein, J. J. (1981). "The Second National Assessment in Mathematics: Area and Volume." *Mathematics Teacher* 74: 704–708.

Johnson-Laird, P. N. (1983). *Mental models: Towards a Cognitive Science of Language, Inference, and Consciousness.* Cambridge, MA.: Harvard University Press, 1983.

Lappan, Glenda, James T. Fey, and Elizabeth Difanis Phillips. *Ruins of Montarek: Spatial Visualization (Connected Mathematics Series.)* Lebanon, Ind.: Dale Seymour Publications, 1998.

National Council of Teachers of Mathematics. *Principles and Standards for School Mathematics.* Reston, Va.: NCTM, 2000.

National Research Council (NRC). *Learning To Think Spatially.* Washington, D.C.: National Academy Press, 2006.

Newcombe, N. S. "Picture This." *American Educator* (2010): 28–35.

Outhred, L., and M. Mitchelmore. "Young Children's Intuitive Understanding of Rectangular Area Measurement." *Journal for Research in Mathematics Education* 31, no. 2 (2000): 144–167.

Sarama, J. and D. H. Clements. *Early Childhood Mathematics Education Research: Learning Trajectories for Young Children.* New York: Routledge, 2009.

Wai, J., D. Lubinski, and C. P. Benbow. "Spatial Ability for STEM Domains: Aligning over 50 Years of Cumulative Psychological Knowledge Solidifies Its Importance." *Journal of Educational Psychology* 101, no. 4 (XXXX): 817–835.

Wheatley, C. L., and G. H. Wheatley. (1979). "Developing Spatial Ability." *Mathematics in School* 8, no. 1 (1979): 10–11.

Wheatley, G. H. *Quick Draw.* Bethany Beach, Del.:Mathematics Learning, 2007.

Winer, Michael. "Fifth Graders' Reasoning on the Enumeration of Cube-Packages in Rectangular Boxes in an Inquiry-Based Classroom." Master's thesis, Ohio State University, 2010.

APPENDIX A

Abbreviated List of the *Common Core State Standards for Mathematics* Standards for Mathematical Practice

Standards for Mathematical Practice

1. ***Make sense of problems and persevere in solving them.***

   a. Seriously attempt to grasp the meaning of a problem.

   b. Analyze givens, constraints, relationships, and the form and meaning of solutions.

   c. Plan a solution pathway rather than simply jumping into a solution attempt.

   d. Consider analogous problems, and try special cases and simpler problems, looking for insight.

   e. Monitor and evaluate progress, and check answers using different methods.

   f. Translate between different representations.

   g. Continually ask, "Does this make sense?"

   h. Understand other approaches.

   i. Draw diagrams of important features and relationships.

2. ***Reason abstractly and quantitatively.***

   a. Make sense of quantities and their relationships.

   b. Decontextualize—abstract a given situation, represent it symbolically, and manipulate symbols without necessarily attending to their referents.

c. Contextualize—pause during the manipulation process to reflect on referents for symbol manipulations.

d. Create a coherent representation; consider the units involved; attend to the meaning of quantities; and know and flexibly use different properties of operations and objects.

3. **Construct viable arguments and critique the reasoning of others.**

   a. Understand and use assumptions, definitions, and previously established results in constructing arguments.

   b. Make conjectures and build a logical progression of statements to explore the truth of conjectures.

   c. Analyze situations by breaking them into cases, and recognize and use counterexamples.

   d. Justify conclusions, communicate them to others, and understand and evaluate the arguments of others.

   e. Reason inductively about data, making plausible arguments that take into account the context from which the data arose.

   f. Compare the effectiveness of plausible arguments, distinguish correct logic or reasoning from that which is flawed, and—if there is a flaw in an argument—explain what it is.

   g. Elementary students can construct arguments using concrete referents such as objects, drawings, diagrams, and actions.

4. **Model with mathematics.**

   a. Apply mathematics to solve real-world problems. [In early grades, this might be as simple as writing an addition equation to describe a situation.]

   b. Make appropriate assumptions and approximations to simplify a complicated situation, realizing that these may need revision.

   c. Identify important quantities in a practical situation and interconnect their relationships.

   d. Interpret mathematical results in the context of the situation and reflect on whether the results make sense.

5. **Use appropriate tools strategically.**

   a. Consider and evaluate the usefulness of various available tools (including technology) when solving a mathematical problem.

   b. Make sound decisions about when various tools might be helpful, recognizing both the insight to be gained and their limitations.

c. Detect possible errors by strategically using estimation and other mathematical knowledge.

d. Use technological tools to explore and deepen understanding of concepts.

6. ***Attend to precision.***

   a. Communicate precisely to others.

   b. Use clear definitions in own reasoning and in discussion with others.

   c. State the meanings of symbols and representations chosen, and use symbols, including the equals sign, consistently and appropriately.

   d. Calculate accurately and efficiently; express numerical answers with a degree of precision appropriate for the problem context.

   e. In the elementary grades, students give carefully formulated explanations to each other. By the time they reach high school they have learned to examine claims and make explicit use of definitions.

7. ***Look for and make use of structure.***

   a. Look closely to discern a pattern or structure (like noticing that arithmetic computations satisfy the commutative or distributive property or that a set of figures all have four sides).

   b. Recognize the significance of an existing line in a geometric figure and use the strategy of drawing an auxiliary line for solving problems.

   c. Step back for an overview and shift perspective.

   d. See complicated things as single objects or as being composed of several objects.

8. ***Look for and express regularity in repeated reasoning.***

   a. Notice if calculations are repeated, and look both for general methods and shortcuts.

   b. Notice regularity.

   c. In problem solving, maintain oversight of the process while attending to the details.

   d. Continually evaluate the reasonableness of intermediate results.

Note: You can access and download this appendix online by visiting NCTM's More4U website (nctm.org/more4u). The access code can be found on the title page of this book.

Appendix B

Abbreviated List of the National Council of Teachers of Mathematics *Principles and Standards for School Mathematics* Process Standards

Process Standards

1. *Problem Solving*

   a. Solve problems using a variety of appropriate strategies while monitoring and reflecting on the problem-solving process.

   b. Build new mathematical knowledge through problem solving.

2. *Reasoning and Proof*

   a. Use various types of reasoning and proof, and recognize reasoning and justification as fundamental to mathematics.

   a. Make and investigate mathematical conjectures.

   a. Develop and evaluate mathematical arguments.

3. *Communication*

   a. Organize, consolidate, and communicate mathematical thinking coherently and clearly.

   b. Analyze and evaluate mathematical thinking and strategies.

   c. Use the language and concepts of mathematics to express mathematical ideas precisely.

4. *Connections*

   a. Interconnect mathematical ideas.

   b. Apply mathematics in contexts outside of mathematics.

5. *Representation*

   a. Create and use representations to organize, reason, and problem solve as well as record and communicate mathematical ideas.

   b. Translate among mathematical representations.

Note: You can access and download this appendix online by visiting NCTM's More4U website (nctm.org/more4u). The access code can be found on the title page of this book.

Appendix C

Pattern-Block Frame Tasks

Use pattern blocks and be sure that the shapes on student sheets exactly match the size of the pattern blocks.

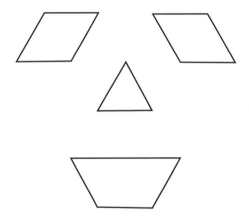

Task Type 1: Shapes not touching

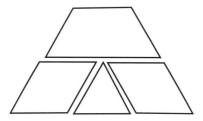

Task Type 2: Shapes not touching but close to suggest another shape

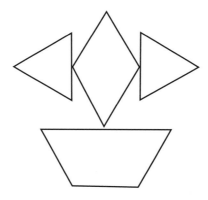

Task Type 3: Shapes touching, but not embedded

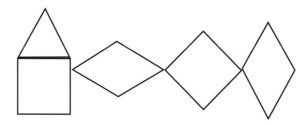

Task Type 4: Shapes touching and shared sides, but not embedded

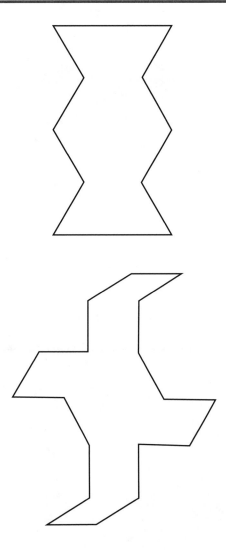

Task Type 5: Shapes embedded, with many sides explicit

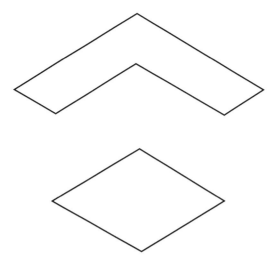

Task Type 6: Shapes embedded, with few sides explicit

Note: You can access and download this appendix online by visiting NCTM's More4U website (nctm.org/more4u). The access code can be found on the title page of this book.

Appendix D

Predict-and-Cover Tasks

Use pattern blocks and be sure that the shapes on student sheets exactly match the size of pattern blocks.

Predict-and-Cover Task 1

How many blocks shaped like this	does it take to cover this shape? Predict; then check with blocks.
	 Prediction _____ Check _____
How many blocks shaped like this	does it take to cover this shape? Predict; then check with blocks.
	 Prediction _____ Check _____
How many blocks shaped like this	does it take to cover this shape? Predict; then check with blocks.
	 Prediction _____ Check _____

Note: You can access and download this appendix online by visiting NCTM's More4U website (nctm.org /more4u). The access code can be found on the title page of this book.

149

Predict-and-Cover Task 2

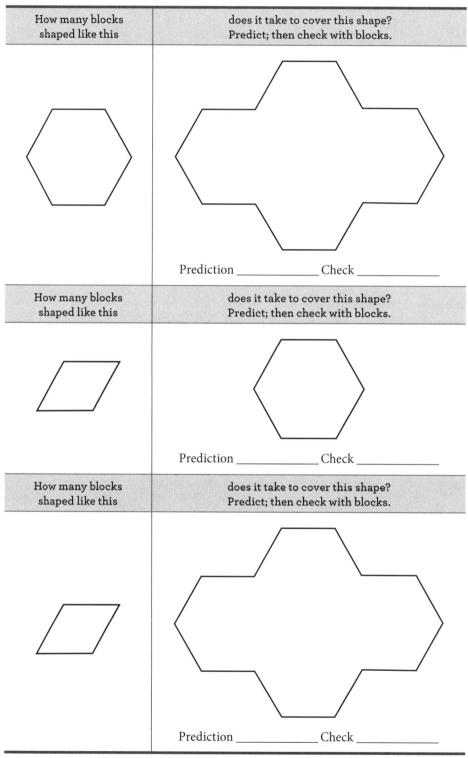

How many blocks shaped like this	does it take to cover this shape? Predict; then check with blocks.
	Prediction _____ Check _____

How many blocks shaped like this	does it take to cover this shape? Predict; then check with blocks.
	Prediction _____ Check _____

How many blocks shaped like this	does it take to cover this shape? Predict; then check with blocks.
	Prediction _____ Check _____

Note: You can access and download this appendix online by visiting NCTM's More4U website (nctm.org /more4u). The access code can be found on the title page of this book.

Reasoning and Sense Making in the Mathematics Classroom: Pre-K–Grade 2

Predict-and-Cover Task 3

Use plastic inch squares and be sure that the squares on student sheets are one-inch squares.

How many shapes like this	does it take to cover this shape? Predict; then check with shapes.
	Prediction _____ Check _____
How many shapes like this	does it take to cover this shape? Predict; then check with shapes.
	Prediction _____ Check _____
How many shapes like this	does it take to cover this shape? Predict; then check with shapes.
	Prediction _____ Check _____

Note: You can access and download this appendix online by visiting NCTM's More4U website (nctm.org /more4u). The access code can be found on the title page of this book.